The Institute of Biology's
Studies in Biology no. 87

Evolution in
Modern Biology

K. J. R. Edwards

B.Sc., M.A., Ph.D.

Lecturer in Genetics, University of Cambridge

Fellow of St John's College, Cambridge

Edward Arnold

First published 1977
by Edward Arnold (Publishers) Limited
25 Hill Street, London W1X 8LL

Board edition ISBN: 0 7131 2650 7
Paper edition ISBN: 0 7131 2651 5

574 EDW

Printed in Great Britain by
The Camelot Press Ltd, Southampton

General Preface to the Series

It is no longer possible for one textbook to cover the whole field of Biology and to remain sufficiently up to date. At the same time teachers and students at school, college or university need to keep abreast of recent trends and know where significant developments are taking place.

To meet the need for this progressive approach the Institute of Biology has for some years sponsored this series of booklets dealing with subjects specially selected by a panel of editors. The enthusiastic acceptance of the series by teachers and students at school, college and university shows the usefulness of the books in providing a clear and up-to-date coverage of topics, particularly in areas of research and changing views.

Among features of the series are the attention given to methods, the inclusion of a selected list of books for further reading and, wherever possible, suggestions for practical work.

Readers' comments will be welcomed by the authors or the Education Officer of the Institute.

1977

The Institute of Biology,
41 Queen's Gate,
London, SW7 5HU

Preface

Since the discovery of the structure of DNA in 1951, molecular biology has developed very rapidly. It has produced a biological revolution in that its ideas have made a considerable impact upon nearly all other branches of the subject.

In evolutionary biology the impact has been primarily through genetics where molecular biology has produced a clear understanding of the nature of the genetic material, of the mechanism by which it replicates, of the way genetic information is coded and of the process of mutation. Of particular importance has been the realization that the sequence of amino acids in a protein is a direct translation of the sequence of bases in RNA, which is, in turn, a direct transcription of the base sequence of DNA. Amino acid sequences and other biochemical properties of proteins in general and enzymes in particular have produced a new set of characters for the study of changes on both long- and short-term evolutionary time scales. The advantage of these characters of proteins over morphological and physiological properties of organisms is that, being much more direct expressions of the genetic information, they give clearer and less ambiguous knowledge of genetic changes in evolution.

In the light of this new information and of the new ideas generated by the explosive growth of molecular biology the time seems right for a brief review of the current ideas of the nature of evolutionary changes and the forces producing them. It is the hope of the author that this book will give some idea of the way in which the modern theory of evolution provides a framework for thinking about many different aspects of biology.

Cambridge, 1977

K. J. R. E.

Contents

1 Theoretical Framework

1.1 Darwin's dilemma

It is now more than 100 years since Charles Darwin and A. R. Wallace read papers on organic evolution to the Linnean Society in 1858; one year later Darwin published *The Origin of Species*. These dates, particularly 1859, are often associated with the birth of the idea of organic evolution. But the idea of evolution had been suggested on a number of occasions during the preceding 100 years or so. What Darwin did was to present an overwhelming weight of evidence for the *fact* of evolution and to suggest a *mechanism* by which it happened. The mechanism was, of course, natural selection and it was around this hypothesis that much of the controversy raged during the next few decades, the fact of evolution being accepted fairly rapidly by most biologists (although rather more slowly by the general public).

Much of the argument about natural selection became inextricably mixed up with the problem of the laws of inheritance. It is axiomatic in the theory of natural selection that evolutionary change will happen only if the differences upon which selection acts are heritable. Thus if in a particular species of birds those individuals with large wings are more likely to survive than those with short, a gradual increase in the wing length of the species over generations will occur only if the difference between long and short wings has a genetic basis. Darwin's views on inheritance were those currently accepted at the time: that inheritance is blending. This interpretation was based upon a general observation that offspring tend, on average, to be intermediate in appearance between the two parents. Analogies have been made between the idea of blending inheritance and the mixing of two fluids of different colour—the 'paint pot' theory. But here is a serious problem for any evolutionary theory, for under blending inheritance individual differences would eventually disappear and all members of a species would become intermediate; how then could natural selection act? To get around this obstacle Darwin was forced to postulate a very rapid rate of creation of new heritable differences, and he accepted the view that adaptations produced in response to particular environments could be inherited. This of course means that he accepted the idea of the inheritance of acquired characters, a concept usually associated with Lamarck rather than Darwin. In fact much of the controversy during the first half of this century over the contribution of the inheritance of acquired characters to evolutionary change is not so much a clash between Lamarckian and Darwinian ideas as between Lamarckian and neo-Darwinian, this latter theory being a modified form of Darwinism incorporating the findings of Mendelian genetics.

Mendelism presents a way out of the dilemma presented by the acceptance of blending inheritance. The essential features of Mendel's interpretation (which were first stated only six years after the publication of *The Origin of Species*, but which went unremarked until 1900) are that while the appearance of offspring may be intermediate between their parents, the basic genetic determinants are particulate, do not blend and can reappear in subsequent generations. Before the incorporation of genetics into Darwinian theory is discussed, alternative views which attempt to account for the diversity found in the organic world should be considered.

1.2   Alternative views and Lamarckism

One view of the great variety of species found is to deny that they were produced by any evolutionary process at all. Such a theory of 'Special Creation' cannot be handled scientifically for it does not allow specific predictions to be made which could then be tested experimentally. Thus while there is no positive evidence in support, it is also very difficult to disprove. Even the order which can be made out of diversity by reconstructing phyllogenetic pathways on the evolutionary assumption could be interpreted as an expression of a 'Great Design' on the part of a Creator. Adherence to an evolutionary theory in preference to a special creation thus depends on the plausibility of the interpretation provided by the evolutionary theory rather than a definitive disproof of special creation.

Of the evolutionary theories, the modern version of Darwinism incorporating Mendelian genetics is now very widely accepted among biologists. But for many years the Lamarckian theory was a powerful competitor and we have seen how Darwin incorporated one element from Lamarck into his own theory, namely the inheritance of acquired characters. The idea that modifications may arise in response to new environmental factors to increase the adaptedness of the individual and that such modifications may then be transmitted to future generations has persisted even until the present day and the era of molecular biology. There is no doubt that organisms do undergo changes in their characteristics in response to environmental alterations (such as plants becoming taller in highly competitive or shaded conditions), what is at issue is whether these environmental induced changes are ever heritable.

A number of experiments have been performed which have been the basis for claims in support of the theory. One of the most famous concerns a toad called *Alytes obstetricians* (the midwife toad). This is unusual in that mating occurs on land not in water, and it differs from closely related species in that the males do not have dark coloured nuptial pads on the palms of the forelimbs; the nuptial pads in the water-mating species are adaptive in that they help the male to grasp the female. At the beginning of this century an Austrian biologist called Paul Kammerer set up some experiments on a laboratory stock of midwife toads in which he forced them to mate in water. He argued that if the inheritance of acquired characters is a tenable theory this experiment would eventually produce males with nuptial pads. And indeed he reported that after a number of generations the males of this stock did develop pads in the breeding season. His claims were greeted with doubt by most Mendelian geneticists, particularly William Bateson, and this scepticism seemed justified when it was discovered in 1926 that the last surviving preserved specimen of a male *Alytes* from Kammerer's water-mating experiment was a fake; the dark patch having been produced by an injection of Indian ink. There now seems to be considerable doubt as to whether Kammerer himself could possibly have perpetrated this particular hoax. In a recent book called *The Case of the Midwife Toad*, KOESTLER (1971) has attempted to reinstate Kammerer's reputation and to reopen the case for the theory of the inheritance of acquired characters. But even if it is accepted that the *Alytes* experiments really did produce males with genuine nuptial pads, this does not provide convincing evidence for the inheritance of acquired characters because there are also plausible explanations based on 'orthodox' genetics. It is now known that males can be found in the wild with nuptial pads; thus the necessary genetic variation does apparently exist in the species to allow the evolution of a high incidence of pads if natural selection should favour this characteristic, as might be expected to happen when forced to mate in water. Thus the possibility exists that Kammerer's original stock was genetically heterogeneous and his results were due to natural selection. Such a selection response might happen even if the starting group of animals showed none with visible nuptial pads, for it is now known that the necessary genetic variation can be present without being expressed because it does not occur in the appropriate combinations of genes. How selection may create these combinations is discussed later (6.2).

Kammerer's results, if genuine, can, like other such experiments, be more plausibly explained by a synthesis of Mendelian genetics and Darwinian natural selection. Such a combination has been referred to several times and most of this booklet will be concerned with an exposition of this modern theory of evolution, which has been called neo-Darwinism or the Synthetic Theory of Evolution.

1.3   Neo-Darwinism

It may be helpful to the reader to list the essential features of neo-Darwinism in order to provide a framework for subsequent chapters. The seven items listed do not exactly match individual chapters but they do indicate the sequence of development.

(a) The fundamental origin of genetic variation is *mutation*. Recent advances in molecular biology have given a much better understanding of the nature of such mutational events and of the factors determining the rates at which they occur.

(b) Given that genetic variation has been produced by mutation, the combinations of genes which actually occur in individuals will be determined by the *breeding system*. Included in the discussion are the differences between sexual and asexual reproduction, and between inbreeding and outbreeding and the effects of linkage between genes on the same chromosome.

(c) If individuals show genetic differences they are very likely to show *phenotypic differences*, but exactly how these are related to their genetic base will depend upon such phenomena as dominance between alleles and interactions in expression between distinct genes.

(d) Certain phenotypes may be better adapted to a particular environment than others. This may happen either because of differences in survival chances or in reproductive potential, but either will lead to a differential contribution to the numbers in the next generation; that is there will be differences in *Darwinian fitness*.

(e) If natural selection does operate changes will occur in the *genetic composition* of the species, the best adapted genotypes replacing the less fit. Rarely a newly arisen mutant may be fitter than the existing form or alternatively the environment may change so that an existing but infrequent and previously disadvantageous mutant becomes the more fit.

(f) Although it might be expected that commonly the most fit type would be markedly predominant much *genetic variation* remains within species. This may be due to the time taken for natural selection to act after environmental changes in time or over space, but there may also be positive selection to maintain genetic differences. An obvious example is the existence of male and female forms, but there are many more subtle forms.

(g) Finally, the heterogeneity of selection within a species may be on such a scale as to lead to the formation of *races* and of *species*.

Underpinning this scheme is an understanding of genetics which has been recently reinforced by the discoveries of molecular biology. It is the purpose of this account to consider critically the extent to which neo-Darwinism provides a satisfactory explanation of the facts of evolution.

2 The Origin of Variation

2.1 Individual differences and their genetic basis

The single most important finding of Mendelian genetics was the particulate nature of the genetic factor or gene. This finding was an inference based upon the detection of certain regular patterns of inheritance shown by specific characters. Thus Mendel inferred from the appearance of a 3 : 1 ratio of tall and short pea plants in the second (F_2) generation following a cross between true-breeding tall and true-breeding short varieties that the difference in height was due to a single factor (later named a gene) which had two alternative forms. Furthermore these two forms retained their integrity even when both were present in a single individual but only one was expressed. In the immediate progeny of the cross (the F_1 generation) all the plants were tall and tallness is dominant to shortness. In the F_2, shortness reappeared in one quarter of the plants showing that the form of the gene (now called an allele) determining shortness had merely remained unexpressed in the F_1. The interpretation of the regular 3 : 1 ratio of dominant: recessive plants was fully supported by further breeding experiments and is encoded as Mendel's Law of Segregation and is a fundamental of Mendelian genetics. The inheritance of height of peas can be represented by the following formal scheme (Fig. 2–1) which could be used as a framework in many other characters in other plants and animals. The letters T and t indicate the two alternative forms (alleles) of the gene controlling height; there are two alleles present in the zygote (diploid phase) but only one in the gamete (haploid stage).

It is not the purpose of this book to reiterate the principles of genetics; the concern here is to emphasize that very rarely is there any direct information about genes, but they can be studied indirectly through the analysis of the inheritance of certain character differences using Mendelian rules. The particular phenotypic differences recognized may be only one of several manifestations of a genetic difference and one of the problems which faces an investigation of natural selection is to find out which, if any, of these effects is (or are) of adaptive significance. Even where there is evidence for the action of natural selection on one particular phenotypic aspect, this may not be the whole story for there may also be other selective effects on further facets of the expression of that genetic difference. A brief treatment of one example may illustrate this point.

It concerns the well-known phenomenon of the rise of industrial melanism. This has been most extensively studied in the peppered moth (*Biston betularia*). FORD (1971) has reviewed the extensive studies on this

6

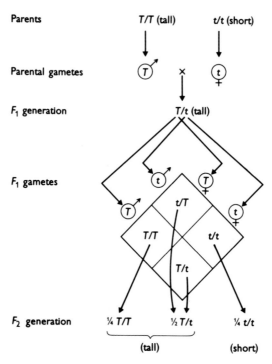

Fig. 2–1 A formal scheme for the inheritance of height in peas. *T* and *t* are the two alleles of the gene determining height. ♂ indicates a male gamete, ♀ a female gamete.

species in which the dark coloured form has become predominant in industrial areas over the last one hundred years or so. Direct experimentation has shown that the dark form is less frequently taken by predatory birds in industrial areas, presumably because it is less obvious than the pale form when resting on tree trunks blackened by soot and with no lichens growing on them. Conversely in rural areas the pale form is the less obvious and suffers less predation. This might lead one to expect that in industrial areas all the moths would be dark and in the open countryside they would all be pale, while mixed populations might occur on the edges of large cities. But this does not seem to have happened and there is always a percentage of the pale form even in the most polluted areas. Ford discusses this maintenance of some genetic variance in terms of possible effects of this gene other than the obvious one on adult colour; and indeed there is experimental evidence to support the view that there are effects on larval viability. The recognizable manifestation of the gene is on adult colour and the dark melanic form is dominant to the pale, that is the heterozygote is dark and is phenotypically indistinguishable from the dark homozygote. Laboratory experiments have demonstrated a greater probability of larval survival for heterozygotes than for homozygotes. As shall be shown later (Chapter 4) a situation in which the heterozygotes have a greater chance of survival than the homozygotes leads to the maintenance of a genetically varied population. For the moment we can consider the melanism case as an illustration of the possible manifold effects of individual genes.

2.2 Mutation

The study of molecular biology has given a much greater understanding of the basis of mutation. That genetic information is coded as the sequence of bases in the double helix of DNA (deoxyribonucleic acid) is now a well known fact. For that part of the information which is translated, via RNA, into proteins the genetic code is such that each amino acid in the protein is determined by a triplet of base pairs in the DNA; thus the sequence of triplets in the gene determines the sequence of amino acids in the protein. All the other properties of the protein, such as enzymatic activity and sensitivity to temperature, stem from this primary sequence. Other genes within the DNA contain sequences which are not translated into proteins but are concerned with the production of RNA molecules involved in the translation mechanism (ribosomal RNA and transfer RNA), with 'punctuation' of translated information, or with other aspects of the regulation of its expression. Much less is known about the specific coding involved in this non-translated information but the base sequence is as important here as it is for those genes which produce proteins.

There are two ways in which the fundamental DNA sequence can be disturbed. Firstly, as a result of a chemical change one of the bases may be transformed into one of the other three bases found in DNA. (The four bases are: adenine which pairs with thymine, cytosine which pairs with guanine.) Secondly, there may be physical changes affecting a sequence larger than a single base pair, such as the addition or deletion of a number of bases. Since the information is read sequentially as triplets even a small addition or deletion can have drastic effects by throwing the 'reading frame' out of phase. Thus the effects on the protein of physical changes in the DNA sequence are likely to be much greater than the effects of a chemical change in a particular base pair which may lead to the substitution of one amino acid for another. But changes of amino acids at certain critical positions in the proteins can have great effects on its function and so produce a detectable phenotypic change. For example sickle-sell anaemia which occurs in many populations in West Africa with an incidence of up to 10% is caused by the substitution of a valine for a glutamine in the β chain of the haemoglobin molecule.

While mutations can be traced back to molecular events in the DNA it is through their phenotypic effects that they are recognized and may be counted. How then is the rate of mutation measured? A rate obviously involves some kind of time scale and the natural unit for measuring mutation is the generation. How the generation is defined depends upon the reproductive system: in bacteria where reproduction occurs by cell division the cell doubling time is the appropriate unit; in higher organisms reproducing sexually the interval between gamete formation in successive generations is used. The reason for the choice of gamete formation as the appropriate phase of the sexual cycle is a practical one: it is possible to design experiments to estimate the proportion of gametes in the output of a known homozygote which carry a mutant allele. Of course the gamete stage can be scored directly only in organisms such as fungi and some algae which have a long lived haploid phase and in a few characters of pollen grains in plants, and so crosses have to be made with appropriate tester genotypes in most organisms.

Suppose that a particular genotypic characteristic is controlled by a gene with two alleles A and a with the normal or 'wild-type' phenotype being dominant and produced by A. Then we could make a cross between the two homozygotes A/A and a/a and we would expect that, barring mutation, the progeny would all be A/a and therefore of wild type appearance. But if a fertilization involved a gamete from the A/A homozygote which carried an allele mutated to a the zygote produced would be a recessive homozygote and therefore distinguishable. Such a situation is schematically represented in Fig. 2–2. Thus the proportion of individuals with recessive phenotypes is a direct measure of the proportion of gametes produced by the A/A parents which carry a freshly mutated a allele. That is, it gives a direct estimate of the mutation rate.

Using this and other methods which are described in many textbooks of genetics such as STRICKBERGER (1976), mutation rates have been estimated for a large number of genes in a variety of organisms. It should be emphasized that these are 'spontaneous' rates with no deliberate

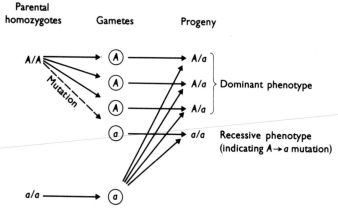

Fig. 2–2 A scheme for estimating the mutation rate of $A \rightarrow a$ by detecting the proportion of a gametes produced by an A/A individual.

treatment aimed at increasing the rate. The use of certain forms of irradiation and certain chemical mutagens can increase the rate for all genes by several orders of magnitude.

The unit of time used in these estimations is determined by the reproductive system, the definition of a generation being different for asexual and sexual reproduction. But it is not only as a factor in the estimation of mutation rates that the reproductive system is relevant to a discussion on the origin of variation. Sexual reproduction permits the formation of new combinations of existing genetic variants, while the extent of such recombination is determined by the linkage relationships of genes and by the mating system.

2.3 Reproductive systems and genetic variation

A vast array of reproductive systems can be found in nature and all that is attempted here is a general classification and a brief consideration of their consequences. The variety of genetic systems is of interest in the study of evolution not only because of effects on the genetic potential for further evolution, but also because the genetic systems are themselves evolutionary phenomena (DARLINGTON, 1958).

2.3.1 Asexual and sexual reproduction

Asexual reproduction occurs by a variety of means over a wide range of organisms: fission in protozoa and bacteria, budding in *Hydra*, vegetative reproduction in many plants, and certain types of parthenogenesis in animals. The essential feature is that the cell division processes underlying all these methods involve a complete replication of all the genetic information in the cell. This is accomplished by mitosis in eucaryotes and by an analogous process in bacteria. Thus the only way in which an offspring may differ from its parent is through the occurrence of mutation.

In sexual reproduction on the other hand the alternating cycle of fertilization and meiosis, illustrated diagrammatically in Fig. 2–3, can produce a reassortment of existing variation between individuals. Fertilization brings together in the diploid zygote genes from different individuals while meiosis produces recombination among these genes of distinct parental origin.

In order to see how these two systems influence the creation of new variation, let us consider a hypothetical, and rather unlikely, example. Suppose that we are considering an organism whose individuals show variation for two genes, for each of which there are two alleles, A, a and B, b respectively, and that there is a population which consists of equal members of two genotypes; A/A, b/b and a/a, B/B. We may then ask the question what are the relative likelihoods of obtaining an a/a b/b homozygote in either sexual or asexual reproduction?

First for sexual reproduction: since there are equal numbers of the two

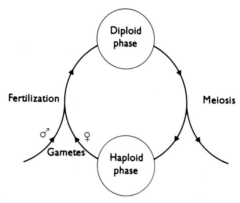

Fig. 2–3 The cycle of sexual reproduction.

genotypes, 50% of the matings will be between the dissimilar genotypes and so 50% of the next generation will consist of the double heterozygote *A/a, B/b* while the remaining half will consist of equal numbers of the two original homozygotes. When the heterozygotes form gametes a quarter will be *ab*, which type will thus constitute an eighth of all the gametes. Since the *a/a, b/b* homozygotes can only be formed when fertilization by two such gametes occurs the frequency in the next generation will be $1/8^2 = 1/64 = 0.015625$ or 1.5625×10^{-2}.

With asexual reproduction the *a/a, b/b* homozygote can only be achieved as a result of a double mutation from either *A/A, b/b* (mutation $A \rightarrow a$) or *a/a B/B* (mutation $B \rightarrow b$). If we assume the mutation rate to be 10^{-5}, which is a fairly representative rate, then the probability of a double mutation is 10^{-10}. But since the homozygote can be achieved by two pathways the probability over a single generation is 2×10^{-10}, and, since for comparison with the sexually reproducing population we have to consider two generations the final rate is 4×10^{-10}.

Thus the frequency of the *a/a, b/b* genotype will be very much higher in a sexually-reproducing population and will be very unlikely to occur at all with asexual reproduction. Similar calculations can be made for other situations and they show that the more genes there are the greater is the advantage of sexual reproduction in generating novel genotypes.

2.3.2 *Variations of sexual reproductive systems*

In the calculations we have just carried out, certain implicit assumptions were made about sexual reproduction. First we assumed that at meiosis in the double heterozygote the various gametic combinations were produced with equal frequency, that is assortment was independent and the two genes were *not linked*. Secondly we assumed that all the gametes formed in the population created a pool from which any combination could be drawn to make a fertilizing pair, that is among the gametes there was *random mating*. Let us now briefly consider these two assumptions, and the effects of relaxing them on the creation of new variants.

Genes are carried on chromosomes and those which are fairly close together are said to be linked. In fact the operational definition of linkage is that in breeding experiments two genes are not inherited independently and do not recombine freely. The extent to which this restriction affects the entire array of genes of an organism depends on two factors: the number of pairs of chromosomes (the haploid number) and the amount of recombination within chromosome pairs. The lower the haploid chromosome number the fewer independent pairs of genes there will be; if there is only one chromosome, as in bacteria, all genes will be linked to each other although the effects on recombination between genes which are far apart will be very slight or even non-existent. That the genes of an organism can become rearranged between chromosomes and that this process can lead to a change in the number of chromosomes can be seen from the study of certain genera such as the plant *Crepis*. In this genus there is variation between species in the haploid chromosome number in the range three to seven. Cytological studies by BABCOCK (1947) as well as morphological comparisons show fairly clearly that the original number was either six or seven and that the species with lower numbers have evolved from this by rearrangement between parts of chromosomes.

Within a chromosome pair there may be great variation in the number of cross-overs (the exchange of material between homologous chromosomes which is the basis of recombination) which may be much more common in certain regions of the chromosome. The genes in a region with very rare crossovers will be inherited as a tightly linked complex and thus certain associations of alleles at different genes may be preserved.

The factors which, at meiosis, control recombination are relevant to future evolutionary possibilities only if heterozygosity occurs at two or more loci simultaneously, and the incidence of such heterozygosity will depend on the mating system. In the earlier discussion on the comparison of asexual and sexual reproduction we assumed that in the latter mating was at random. The most common deviation is that mating is much more likely to be between individuals which happen to be in close proximity. If the movements of organisms between fertilization and reproductive maturity are small, individuals which are near are likely to be related. Thus some increase of mating between relatives, that is of inbreeding, will probably occur. The effects of this on the frequency of heterozygosity can be most easily seen if we consider the case of the most extreme form of inbreeding, namely self-fertilization. In self-fertilization the uniting gametes are both derived from the same diploid individual and this can happen only in hermaphroditic organisms, such as many 'lower' animals and most plants.

Let us suppose that we start with an F_2 generation with the proportions of the three genotypes A/A, A/a and a/a being $\frac{1}{4}$, $\frac{1}{2}$, $\frac{1}{4}$. If this population now reproduces by random mating the genotypic proportions will remain as before and the frequency of heterozygosity will be 50%. But if reproduction is entirely through self-fertilization the heterozygotes will produce a $\frac{1}{4}$, $\frac{1}{2}$, $\frac{1}{4}$ segregation but the homozygotes will breed true, producing more identical homozygotes. Thus the next generation will contain only 25% heterozygotes and at each further generation of self-fertilization the proportion will again be reduced by half. If two genes are being considered simultaneously the relative effect of self-fertilization is to reduce the proportion of double heterozygotes to a quarter of that expected under random mating. Thus the opportunity for recombination is severely restricted under self-fertilization, and is restricted to a lesser extent under any other type of close inbreeding.

In nature we find a number of mechanisms which prevent self-fertilization. An obvious one is sexual differentiation so that each individual produces only male or only female gametes. Most commonly the switch mechanism is due to the segregation of sex chromosomes so that the possession of two identical sex chromosomes, X/X, leads to femaleness, while males are produced by the X/Y chromosome configuration. But this is not universal, for in birds and in Lepidoptera (moths and butterflies) males are X/X and females X/Y, while in some Hymenoptera (bees and wasps) males develop from haploid, unfertilized eggs and females from diploid, fertilized eggs. Most plants are hermaphrodite, but many are self-incompatible with the pollen from a given plant being unable to germinate or grow on the stigma of the same plant and there are certain other barriers to self-fertilization.

Devices such as sex-determining mechanisms and self-incompatibility systems have evolved to prevent self-fertilization, but they do not restrict other forms of inbreeding such as brother-sister mating (although most human societies have developed social or religious taboos against matings between close relatives). In fact most species probably inbreed to some extent and this, together with linkage, will tend to restrict the free recombination of the available genetic variation. It is not easy to say whether this is likely to be a 'good thing' or a 'bad thing'. In very stable environments where the best adapted genetic combinations probably already exist in high frequency it will be a disadvantage to have a great deal of recombination, but in rapidly changing circumstances the opposite may well be true if a new force of natural selection favours genotypes which have previously been non-existent or very rare.

3 Natural Selection in Action

3.1 Darwinian fitness

Natural selection is the essence of Darwin's theory. We shall consider later whether it is an adequate cornerstone to support the theory of biological evolution; here the ways in which natural selection can produce genetic changes will be examined together with a few examples of studies of it in action.

Fundamental to the treatment is the concept that some individuals are genetically better adapted than others and are then said to be more fit. The phrase 'survival of the fittest' is commonly associated with Darwin (and in fact appeared in the heading for one of the chapters of *The Origin*), but the term Darwinian fitness is in fact concerned with more than survival, for an individual with a good chance of surviving to adulthood but which is then infertile and makes no contribution to evolution cannot be said to be fit in the Darwinian sense. The important measure is the number of offspring contributed to the next generation relative to that of other individuals of different genotype. The term fitness refers to this contribution, and it is convenient to classify variations into those affecting the chances of survival from zygote formation to reproductive maturity (viability), and those affecting the actual numbers of offspring produced by mature adults (fertility).

Evolution occurs only if differences in fitness have a genetic basis, but how can we establish that individuals with differing fitness are genetically different? The entire genotypes of the two individuals cannot be known, only small parts, namely the condition of a limited number of genes recognizable because of known effects on certain characters, although these detectable manifestations may or may not have a direct effect on fitness. On the other hand, we may measure directly some character which we assume must make a strong contribution to overall fitness, such as the number of eggs laid by birds. If we know from breeding experiments that clutch size is a heritable character we may then expect that natural selection will lead to an increase in the frequency of those genes which increase the clutch size. The weakness in this line of approach is that whatever character we choose is likely to be only part of the story. LACK (1948) has shown that in the starling there is optimum clutch size, above which the food supply per chick is limiting and their viability becomes so low that the relative contribution of these larger clutch sizes is actually lower (Fig. 3–1).

One other way in which we study the effects of natural selection and which has the advantage that quantitative estimates and predictions can be made, is to study the changes in frequencies of genes whose segregation

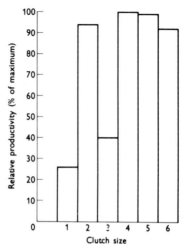

Fig. 3–1 Relative productivity of various clutch sizes in starlings (LACK, 1948). The total number per nest of young surviving to 3 months after fledging for a clutch size of 4 is set at 100. All other values are expressed as a percentage of this.

can be readily detected on some easily observable character. Before we consider a few specific cases, let us turn to a brief theoretical treatment to see how natural selection might operate. We can do this by the use of some very simple algebra. Firstly, however, it is necessary to define the concepts of allele and genotype frequencies.

Again we consider a hypothetical gene with two alleles which we shall call A and a. A population of individuals will consist of the three genotypes A/A, a/a and A/a and we can designate their relative proportion P, Q and R. These are the *genotype frequencies*. Since these are proportions, $P + Q + R = 1$. We can also describe the genetic composition of the population in terms of the relative proportions of A and a alleles and we can call these proportions p and q. These are *allele frequencies* (often called *gene frequencies*). Since all the alleles in A/A individuals and half those in A/a are A we can easily see that

$$p = P + \tfrac{1}{2} R$$
Similarly $\quad q = Q + \tfrac{1}{2} R$

Finally we can consider the relationships between p and q on the one hand and P, Q and R on the other. If mating is at random this means that the gametes, which will be in the proportions p and q carrying the A and a alleles respectively, fuse in random pairs. Thus the proportion of A/A zygotes formed will be p^2, of a/a will be q^2, while the heterozygotes A/a will be 2pq. Thus with random mating we have

$$P = p^2$$
$$Q = q^2$$
$$R = 2pq$$

This relationship between gene and genotypic proportions is called the Hardy–Weinberg Law being discovered independently but coincidentally by G. H. Hardy and W. Weinberg in 1908. Both gene and genotypic frequencies remain constant from generation to generation providing that mating is always at random and that there are no disturbing features such as selection.

We can now see what happens if the three genotypes do not have identical fitnesses. Let us assume that individuals of a/a genotype have a 20% lower chance than the other genotypes of surviving to reproductive maturity. Since we are concerned with relative fitnesses we can put the A/a and A/a fitnesses as 1.0 while a/a has a fitness of 0.8.

Genotypes	A/A	A/a	a/a
Relative fitnesses	1.0	1.0	0.8
Initial proportions	p^2	$2pq$	q^2
Proportions at maturity	$p^2/(1-0.2q^2)$	$2pq/(1-0.2q^2)$	$0.8q^2/(1-0.2q^2)$

The initial proportions at zygote formation before the differential effects of selection have operated are in Hardy–Weinberg agreement; but after selection the relative proportions have become $p^2 : 2pq : 0.8q^2$ since fewer of the a/a genotype have survived. As these values do not sum to 1.0 but to $(1-0.2q^2)$ we must divide each term by the latter to bring them all to true proportions. Thus the gene frequency has changed and the new value

$$p' = \frac{p^2+pq}{(1-0.2q^2)} = \frac{p(p+q)}{(1-0.2q^2)}$$

and since $p+q=1$

$$p' = \frac{p}{(1-0.2q^2)}$$

Thus the change in $p(\varDelta p)$

$$\varDelta p = p' - p = \frac{p}{(1-0.2q^2)} - p$$

$$= \frac{p - p(1-0.2q^2)}{(1-0.2q^2)} = \frac{p - p + 0.2pq^2}{(1-0.2q^2)}$$

$$= \frac{+0.2pq^2}{(1-0.2q^2)}$$

Thus

$$\varDelta q = \frac{-0.2pq^2}{(1-0.2q^2)}$$

We can generalize this treatment by noting that 0.2 is the relative disadvantage of a/a. We can call this s and thus

$$\varDelta q = \frac{-spq^2}{(1-sq^2)}$$

A few moments consideration of this equation will show that the change in allele frequency between successive generations is determined not only by s (the selective differential) but also by the existing values of p and q. In particular, as q gets very small, Δq will also fall. Thus if s=0.2 and p=q=0.5, Δq=0.0263. But if p=8 and q=2, Δq=0.0064. Selection against a recessive allele becomes very ineffective as the allele becomes rare, a fact of considerable importance because most deleterious mutations which occur are recessive. The relationship between q and Δq for certain values of s is shown in Fig. 3.2.

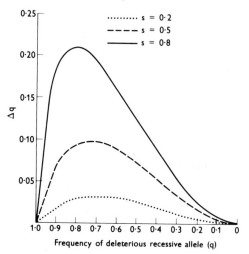

Fig. 3–2   The relationship between the change of allele frequency per generation (Δq) and the allele frequency (q) for 3 different selection intensities.

Similar calculations may be made for situations other than selection against an allele which is completely recessive. The reader may like to satisfy himself, for example, that for selection against a dominant, where the relative fitnesses would be:

Genotype	A/A	A/a	a/a
Relative fitnesses	1 − s	1 − s	1

$$\Delta q = \frac{+ sq^2(1-q)}{1-s(1-q^2)}$$

3.2   The case history of industrial melanism

The occurrence of a high frequency of dark coloured forms in industrial areas has been noted in many species of moths, but the best studied case is the peppered moth *Biston betularia*. Detailed accounts of the observational and experimental investigations can be found in a number of books, such as FORD (1971). A brief description of the main findings follows and is used to illustrate the methods used in investigating natural selection.

Before about 1850 all *Biston betularia* moths were white with black speckling on wings and body. A black variety was first caught in Manchester in 1850. The black form is called 'carbonaria' while the pale is known as 'typica'. During the second half of the last century the frequency of the carbonaria form increased rapidly until it accounted for over 90% of the Manchester populations. Similar changes happened in other large industrial cities, but in rural areas the melanics remained unknown or were rare. Figure 3–3 shows the recent distribution of melanics in Britain and while it is obvious that the industrial areas have a very high frequency of dark forms, these are also surprisingly frequent in many rural areas in the east (for example, East Anglia). Rural areas to the west, which is upwind since the prevailing winds are westerlies, have very few melanics.

In order to understand this evolutionary process we first need to know the genetic basis of the observed phenotypic differences. It is fairly simple: the difference between carbonaria and typica is due to the segregation of two alleles of a single gene with the dark form being completely dominant. There is also another melanic form, insularia, which, as can be seen from Fig. 3–3, is generally not at all common except in the south and west. The insularia phenotype is intermediate between carbonaria and typica and is produced by a third allele of the same gene. Insularia is dominant to typica but recessive to carbonaria. Thus if we call the three alleles C^c, C^i and c, for carbonaria, insularia and typica respectively, we can see that there are the following relationships between phenotypes and genotypes:

Phenotypes	Genotypes
Carbonaria	C^c/C^c; C^c/C^i; C^c/c
Insularia	C^i/C^i; C^i/c
Typica	c/c

Most of the remaining discussion will deal only with the C^c and c alleles, but the principles could be applied to C^i.

The observed rise of melanism is correlated with industrial pollution and it seems likely that natural selection favours the melanic forms in industrial areas because their colour gives them protection from predation by birds while the converse holds in rural areas. A number of experiments have demonstrated that this in fact happens. In one such set of experiments KETTLEWELL (1956) released equal numbers of marked typicals and carbonaria into a wood near Birmingham. The released moths were individually marked so that if recaptured they could be recognized. Kettlewell found that when he trapped the moths a few days later many fewer of the typicals than of the carbonaria were recaptured and presumably fewer had survived. The converse results were obtained when equal numbers of the two types were released in an unpolluted Dorset wood. The actual results are given in Table 1 and show that in each case the chances of survival of the more obvious form are only about one third to one half of those of the cryptically coloured form. Direct

18

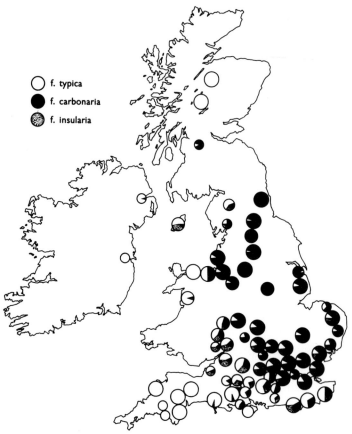

Fig. 3–3 Distribution of melanic and non-melanic forms of *Biston betularia* in U.K. (Reproduced, with permission, from KETTLEWELL, 1958.)

observations of predation by birds upon moths resting on tree trunks have supported these findings.

Direct experimentation has shown that natural selection through differential predation acts in such a way as to favour the melanic allele in industrial areas. But is it sufficient to account for the observed rate of increase of the allele following the industrial revolution? To answer this question we can make some calculations which are the converse of those developed in the first section of this chapter. Then we had assumed certain selection differentials and had predicted changes in gene frequency; now, we have certain observed changes in frequency, and need to estimate the selection differential. The details of the calculations need not concern us, but HALDANE (1924) estimated the selective disadvantage of the typical form in the Manchester area for the period 1848–1898. Since *Biston betularia* has one generation per year, this period represents fifty generations. The average selective disadvantage over the period was estimated at 30%. This is the value required to increase the frequency of the carbonaria allele from 1% to 99% over fifty generations and it is similar to the estimates obtained from Kettlewell's experiments.

Since both the analysis of frequency changes and direct experimentation show a clear selective advantage for melanism in industrial areas one would expect that the typical form would eventually disappear altogether. But this has not happened. By 1900 the typical phenotype had declined to less than 5% in Manchester but since that time it has not been reduced further, and in all industrial areas the spread of melanics in *Biston* seems to have plateaued at 80–90%. Why is this? The answer is that it probably represents an equilibrium due to a balance of conflicting selective forces, although exactly what these are in the case of industrial melanism is still not known. A much better understood example of such a balance is discussed in the next section.

Table 1 Comparison of recapture figures for released, marked carbonaria and typica moths in (a) a rural unpolluted area, and (b) urban polluted areas

Area		Carbonaria	Typica
(a) Unpolluted	Released	473	496
(Dorset)	Recaptured	30	62
	% Recapture	6.3	12.5
(b) Polluted	Released	601	201
(Birmingham)	Recaptured	205	32
	% Recapture	34.1	15.9

3.3   Sickle cell anaemia and malaria

Sickle cell anaemia is a heritable disease of man found particularly in populations in many parts of Africa, in certain Mediterranean countries, and in India. The genetic basis is simple for it is due to the segregation of a single gene. Homozygotes for the mutant allele suffer from severe anaemia which is usually lethal before adulthood is reached, while heterozygotes are normal except that their red blood cells become sickle shaped if blood samples are subject to low oxygen tension. In the homozygotes the sickle-shaped red blood cells occur under normal oxygen conditions. The immediate biochemical expression of this gene is the production of the β chain of adult haemoglobin. Thus the allele which determines normal adult haemoglobin is designated Hb^A, while that producing the abnormal form associated with sickling is called Hb^S. In the Hb^S/Hb^S homozygote only abnormal haemoglobin occurs and severe anaemia results, while in the Hb^A/Hb^S heterozygote both forms are produced (one by each allele) and the red blood cells can function normally under most conditions.

Thus the Hb^S allele appears to be a seriously deleterious recessive and it has been estimated that the probability of an Hb^S/Hb^S homozygote surviving to maturity is only about 20% of that of a dominant genotype. Why then is the allele so common in certain parts of the world, its frequency being 10–20% in many parts of West Africa and locally reaching 40%? It was A. C. Allison who noticed that the distribution of the sickle cell gene paralleled that of falciparum malaria (Fig. 3–4), caused by the blood parasite, *Plasmodium falciparum*. Could it be that this sickle cell allele in some way increased resistance to malaria, and could this be a selective force to balance the disadvantage of anaemia? Let us consider some of the possibilities.

If the homozygote Hb^S/Hb^S were more resistant to malaria than the other genotypes this would, of course, tend to counteract the anaemic effect. But while there is a minute probability that the two selective forces would exactly balance it is much more likely that there would be a net imbalance one way or the other. And if that happens there would be a steady selection towards uniformity for either one allele or the other. In view of the known high lethality of the anaemia effect it seems unlikely that the malaria effect could more than counterbalance it and therefore it remains a puzzle to explain the high incidence of the Hb^S allele.

In the argument in the preceding paragraph we have assumed that the malaria resistance effect we have postulated must be recessive. But this is not the only possibility and when we remember that the Hb^A/Hb^S heterozygote produces both kinds of haemoglobin it seems reasonable to consider some effect in the heterozygote. In fact ALLISON (1964) has collected evidence to support the view that the 'carriers' of the sickling trait, that is the heterozygotes, are less likely to contract a serious attack of malaria and are less likely to die from the disease than are the Hb^A/Hb^A homozygotes. In heterozygotes it has been found that infected individuals have only about one third the number of parasites in their blood

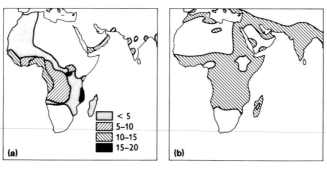

Fig. 3–4 Distribution in Africa and South Asia of (a) sickle cell allele; and (b) malaria. In (a) the key indicates the frequency of the sickle cell allele as percentages. (Reproduced, with permission, in a modified form from CAVALLI–SFORZA and BODMER, 1971, ALLISON, 1964 and BOYD, 1949.)

compared to infected homozygotes. Furthermore of a sample of about 100 children who died because of malaria it was found on autopsy that only one was a heterozygote, although the incidence of heterozygotes in the population at large was about 20%.

Thus we have a situation where the heterozygote has a better chance of survival than the sickling homozygote because it does not suffer from anaemia, and also a better chance of survival than the normal homozygote because it has greater resistance to malaria. The heterozygote is then the most fit, a situation for which the terms heterosis or heterozygote advantage are used. In fact ALLISON (1964) has attempted to quantify these differences in relative survivals by comparing the frequencies of the genotypes in a sample of very young children with those in a sample of mature adults. The calculations are analogous to those developed in the first section of this chapter for selection against a recessive and the relative values Allison obtains are:

Genotypes	Hb^A/Hb^A	Hb^A/Hb^S	Hb^S/Hb^S
Survival values	0.85	1.00	0.20

Since these are relative values the heterozygote, having the highest value, is taken as a standard and its survival value set 1.0. If we assume that all those individuals who reach adulthood have an equal chance of reproducing whatever the genotype, we can regard these values as estimates of relative fitnesses, that is of relative contributions to the total number of genes in the next generation.

Since the heterozygote is the most fit genotype and since heterozygotes are not true breeding but will always produce some homozygotes in their progeny we would not expect to find populations being genetically homogeneous. Let us see what will happen to the gene frequencies in successive generations. We will start with a frequency of the sickle cell allele of 0.16, thus the normal allele must have a frequency of 0.84. From the Hardy–Weinberg Law discussed at the beginning of this chapter we expect the frequencies of the three genotypes Hb^A/Hb^A, Hb^A/Hb^S, and Hb^S/Hb^S to be p^2, $2pq$ and q^2 respectively where p and q are the allele frequencies. Thus we can produce the following scheme, working in exactly the same way as the earlier example:

Genotype	Hb^A/Hb^A	Hb^A/Hb^S	Hb^S/Hb^S
Relative fitness	0.85	1.00	0.20
Initial proportions (assuming q=0.16)	0.706	0.269	0.026
Proportions at maturity	0.600/0.874	0.269/0.874	0.005/0.874

Thus the new frequency of the HB^S allele is

$$q' = \frac{0.005 + 0.135}{0.874} = 0.16$$

So the allele frequency has not changed and is at an equilibrium.

Such an equilibrium is common in those parts of the world where malaria is endemic and the frequency of the Hb^S allele ranges from 0.10–0.20. When malaria is eradicated by control of the mosquito vectors one would expect the frequency of the sickle cell allele to decline steadily because selection is now in one direction only. In no area has such a control been completed for long enough for changes yet to be detectable, but there is one 'experiment' where changes can be observed. This is the decline of the frequency in American Negro populations who have not been subjected to malaria for a period of about 200–300 years. Here the frequency of the sickle cell allele is about 0.05, while in those parts of West Africa from which most of the American Negro population were derived the frequency is about 0.10.

The projected decline in the frequency of sickle cell anaemia where malaria is eradicated and the rise in melanism due to idustrialization are both evolutionary changes due to man's activities. But, of course, the activities which produced these changes were not designed to do so; that they caused new selective forces was incidental. In some cases, however, man has deliberately set out to influence certain species and has produced a change in the forces of natural selection. This has happened wherever attempts have been made to control, and ideally to eradicate, pests and diseases, or to improve crops and stock under domestication.

3.4 Evolution of resistance to control measures in pests and pathogens

A great deal of publicity has been given to the evolution of strains of pathogenic bacteria resistant to commonly used antibiotics. In an attempt to limit this spread there has been encouragement for a careful use of, say, penicillin and for the use of alternatives, possibly in a cyclic way over time. However, the discovery of genetic factors which can confer multiple drug resistance and can be transmitted between species has made the problem more serious.

The idea that control measures may be used in a cyclic pattern if genetic resistance begins to appear rests upon the assumption that the mutation producing the resistance will actually be disadvantageous in the absence of the control. An illustration of this occurs in the resistance to warfarin, an anticoagulant, which has arisen in some rat populations. The gene for warfarin resistance is disadvantageous in the absence of the poison because it produces a high requirement for vitamin K.

4 Maintenance of Variation

4.1 Populations and polymorphisms

In the previous discussion on natural selection frequent reference was made to the genetic composition of populations. But how do we define a population in the context of evolutionary changes? The answer lies in terms of the concept of the mating system considered in Chapter 2. A genetic population consists of those individuals from whom the mating pairs will be drawn. It is an interbreeding group and is sometimes called a Mendelian population, a name which emphasizes the genetic criterion used in defining it. We have seen how random mating can produce proportions of the various genotypes in a particular relationship called the Hardy–Weinberg Law. To mate at random means that the probability of getting a cross between two genotypes depends solely on their relative frequencies, that is the members of a mating pair are drawn entirely at random from the population.

The concept of a genetic population as an interbreeding group does not, however, depend on the mating being random. It could also hold under a system of selection of mates of similar phenotype (and therefore, to some extent, of similar genotype). This is called positive assortative mating and a few moments consideration will show that it leads to genotypic proportions which differ from Hardy–Weinberg expectations. The converse situation when mates are chosen because they are phenotypically dissimilar is called negative assortative mating. In fact an interbreeding group occurs in all situations where there is some mating between individuals; it would have no meaning only for organisms which reproduced solely through asexual methods or which were completely inbreeding because they were obligate self-fertilizers. It is very unlikely that such organisms exist, but a few may so nearly approach these states that it is doubtful if the concept of a genetic population can be usefully applied to them. Cultivated forms of barley, for example, produce over 99% of their seed through self-fertilization. But cultivated barley is not a naturally occurring organism and while there are a few wild plants which show very high proportions of self-fertilization, the great majority of natural species have much lower levels of inbreeding. Thus we can generally apply the term population in a genetic sense as a basic unit of evolutionary change, although there are, of course, problems of precise delimitation in any specific case.

In practice we rarely have much information on the mating system and we often use ecologically based criteria to define a population. Thus we regard all the gulls of a certain species inhabiting (that is nesting on) a small island as being members of a single interbreeding population,

although it may be that the inhabitants of an adjacent island also belong
to the same population. With non-mobile organisms, such as oak trees in
a wood, it is easy to identify the members of a population but special
problems arise with migratory animals. Many species of ducks and geese
migrate between small summer breeding colonies and large winter
feeding colonies, the latter being composites of several summer groups.
But which is the Mendelian population? For geese, mating pairs are fairly
permanent and tend to be formed from within the summer colonies,
while ducks form less persistent pairs in the winter feeding groups. Thus
the effective Mendelian populations are the summer colonies in geese but
the larger winter groups for ducks.

   Despite these difficulties an empirical approach can be made by taking
a sample of individuals which can be regarded as representative
(providing the sample is large) of some natural population. The genotype
frequencies of the sample can be measured and compared with other
samples taken from what appear, on the available evidence, to be
different populations. It may be that in some comparisons between
populations of a single species a particular gene may be uniformly
homozygous for one allele in the first population but for a different
allele in the second. But more often we find situations in which the same
alleles are present in both populations but in different proportions. In
fact one of the most important, and surprising, findings in evolutionary
biology in the last decade has been the very high levels of genetic
heterogeneity which have been discovered in populations of a wide range
of different organisms. This discovery has followed from the use of
techniques for detecting variants in proteins generally and enzymes
particularly. The most important technique has been gel electrophoresis
which picks up differences in the net electrical charge or in the size of a
protein. Electrophoresis is a fairly simple technique which can be applied
to large samples and so is particularly useful for population studies. What
has come out of such surveys is that, over many species, an average of one
third of all genes producing proteins will be genetically heterogeneous in
any particular population. The exact number of genes in any organism,
except a few extremely simple viruses, is not known but it is thought to be
of the order of 5000 in that classic genetic organism *Drosophila
melanogaster*. Thus in a single population we would expect 1500 or more of
these genes to be represented by two or more different alleles. Making the
conservative assumption that there are two alleles at any one gene then
there are three possible genotypes, and for 1500 genes there would be
3^{1500} possible combinations! This enormous number is equivalent to
10^{715} and is not only far larger than any conceivable population but is
even much larger than one estimate of the number of particles in the
universe, which is 10^{78}!

   Such an enormous reservoir of genetic variation provides the basis of
future evolutionary response to the demands of natural selection, and the
question arises: what forces are responsible for maintaining the co-

existence of two or more alleles of each gene? As we have seen earlier the basic origin must be mutation and mutation is recurrent. But usually the recurrent mutant allele is deleterious so we might expect that its frequency in the population would be low because mutation rates are generally low and could be balanced by a relatively small selective disadvantage. In fact such a balance is rarely likely to produce a frequency over 1% and the estimates of about one third of all loci being heterogeneous exclude all loci which have one very common allele and a second very rare allele with a frequency of less than 1%. Populations in which, for one particular gene, there are two or more alleles with frequencies over 1% are said to be genetically polymorphic, a concept developed by E. B. Ford and defined formally in FORD (1971). Since recurrent mutation to an unfavourable allele cannot maintain a polymorphism, we are left with two possibilities: either there is some balance of forces of selection or the alleles are equally well adapted and there is no selection. Considering selection first, I am not suggesting that a population is selected to be variable *per se*, but that there may be situations where the different genotypes vary in fitness in such a way that no one allele can become fixed in the population. We have considered one such possibility already when discussing the maintenance of the sickle cell gene in West Africa; the heterozygote has a superior fitness to both homozygotes and because heterozygotes cannot breed true the population remains polymorphic.

Ford gives details of a number of examples of such heterozygous advantages maintaining a polymorphism and also discusses other mechanisms. One of these is called frequency-dependent selection and it can best be explained by a simple example. Predators, such as birds, learn to associate certain patterns and shapes with a source of palatable food. Thus a mutation in some aspect of the pattern or colour of a prey species might have an advantage because the predators failed to recognize it as food. If the mutant suffered less predation, its frequency in the population would increase, but this would be accompanied by a greater likelihood that the predators would learn that it too was palatable. In other words the rarer the allele the higher its fitness and this kind of frequency-dependent selection would maintain a polymorphism.

4.2   Are all differences adaptive?

If there are no selective forces acting upon the various alleles in the population then, by definition, they are of equal fitness. What then will produce changes in gene frequency? Every population is of finite size and many may be relatively small; thus chance deviations can play a part and gene frequencies may fluctuate from generation to generation over time or between populations over space.

Such differences are generated by sampling effects and are essentially unpredictable in any particular case. However, we can estimate the frequency with which certain changes will occur for a certain population size using probability theory. The process is an extension of the methods used in estimating the probability of a particular outcome in a coin-spinning exercise or of a particular hand of cards in bridge. In order to illustrate its application to gene frequency changes let us consider a single locus with two alleles which are initially at equal frequency. In the next generation the gene frequency may have changed in either direction. Fig. 4–1 shows the distribution of the probabilities of various gene frequency outcomes for population sizes of 5, 10 and 100. The histograms show these probabilities expressed as percentages for the gene frequency classes.

The comparison of these figures shows clearly that chance deviations from one generation to the next (random genetic drift this process has been called) is much more likely to happen if the population has only a few individuals. We can easily extend this argument to differences between populations if we note that the range of possible outcomes is the range of gene frequencies which could exist in a series of populations of limited size all derived from the same ancestral population.

While such calculations are useful in showing what changes might happen due to finite population size alone, the problems which arise in practice are of trying to decide whether an observed difference between populations, whether over time or space, is due to natural selection or to drift or to some combination of the two. The argument over this issue, or to be more precise over their relative contributions in producing genetic changes, has been going on for years amongst evolutionists and has recently developed a new vigour because of the discoveries of protein variation referred to in the previous section.

The identification of the alternative forms of a particular enzyme by electrophoretic techniques depends upon the various forms retaining enzymatic activity, for they are visualized as bands on a gel by treatment with a suitable substance plus a dye which will form an insoluble stain with the product of the enzymatic reaction. It has been argued by a number of geneticists that the genetic variation uncovered in this way is irrelevant to selection because alternative forms are all physiologically functional. But of course this may not mean that they are *equally* physiologically functional in all conditions under which they might have to operate in the organism. For example, if two such allelic variants produced forms of an enzyme which had different optimum temperatures, it would be obviously worth while attempting to correlate the distribution of these two forms in nature with environmental temperature ranges.

Meanwhile the controversy over the role of random genetic drift in evolutionary change continues and we will leave it with this one consideration: random drift cannot maintain a polymorphism indefinitely. While the gene frequencies of adaptively neutral alleles may bounce backwards and forwards irregularly sooner or later one allele will

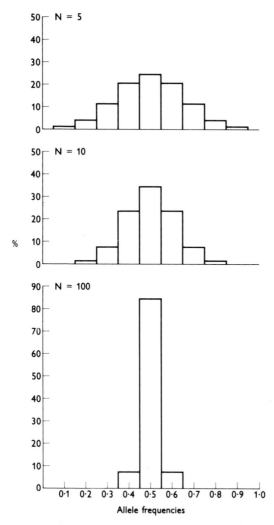

Fig. 4–1 Frequency distributions of possible outcomes after one generation, starting with allele frequencies equal, when only 5, 10 or 100 individuals breed.

be completely lost from the population and once this has happened genetic variation is irretrievably lost unless mutation shall recreate it. Such loss is more likely to happen in small populations which we therefore expect to be genetically more homogeneous, as indeed we generally find to be the case. As long as such alleles remain adaptively neutral then the loss of one does not matter, but if conditions alter so that selection begins to operate at this gene some evolutionary potential will have been lost.

4.3   Long-term survival and hidden variation

The chances that a population will be represented by its descendents many generations hence will depend upon (i) its initial genetic composition being such as to ensure that it produces enough offspring to replace itself in the short term, and (ii) it having sufficient genetic variation to respond to any change in conditions which produce novel selective forces in the long term. We have considered how genetic variation may be maintained by certain balances of selective forces but will tend to be dissipated by random genetic drift in small populations. Current evidence suggests that most populations contain an enormous reservoir of genetic differences, certainly in so far as they produce differences in properties of enzymes and other proteins. In fact we have seen that the number of possible genetic combinations is likely to be so huge that only a very small proportion of them will actually occur in even a large population. If the various combinations were also formed through random associations of the component genes then their probabilities of occurrence would depend solely on the frequencies of the alleles concerned. Thus, if the frequency of individuals of genotype *A/A* were 50% and of *B/B* also 50%, the expected frequency of *A/A B/B* would be 25% and this genotype would almost certainly be found in any population. On the other hand if the component frequencies were both 1%, that of the *A/A B/B* genotype would be 0.01% and it would be likely to be absent completely from most populations except very large ones. But should the environmental conditions now change so that *A/A* and *B/B* become favoured by natural selection their frequencies would rise as would the probability that *A/A B/B* would occur, and so a new genotype would arise.

Such calculations take no account of the biology of the organism, a number of components of which may greatly influence the outcome. One of these is linkage. If in the example used above the two genes had been linked such that the genotypes *Ab/Ab* and *aB/aB* were much more common than their reciprocals (but both of course being much less common than *ab/ab*) then the probability of finding *AB/AB* would be much less than before and its appearance during the evolutionary change we have postulated would be much delayed. Indeed if the initial linkage associations were complete (that is, the genotype *AB/AB* did not occur at all) the spurt of evolution dependent upon the appearance of *AB/AB* genotypes would depend on a recombinational event.

Many laboratory experiments using artificial selection have produced results which suggest that such a sequence of events might well have happened. The most favoured organism for such experiments has been *Drosophila melanogaster* because it can breed so quickly, passing through a generation in about two weeks in the most favourable conditions; and frequently the character on which selection has been practised has been the number of hairs (chaetae) on certain parts of the body. Consider for example an experiment carried out by MATHER (1941) on the number of hairs on the ventral surfaces of the fourth and fifth abdominal segments. Mather crossed two inbred laboratory lines with very similar average hair numbers, and produced an F_2 generation from which he began to select two distinct lines, one for high number of hairs and one for low. In each generation of the high line he selected those flies with the highest number of hairs and used these as the parents of the next generation. The results for this high line are shown in Fig. 4–2. The mean numbers for the two parental lines were 36.1 and 39.9 hairs respectively while the F_2 generation had a mean of 38.0, this latter value being regarded as a standard and indicated by a horizontal line on the selection graphs.

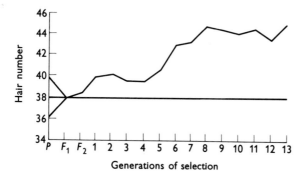

Fig. 4–2   Response to continued selection for high abdominal hair number in a *Drosophila melanogaster* population following a cross between two laboratory inbred lines. (Reproduced, with permission, in a modified form from MATHER, 1941.)

The results show that there is an immediately small response up to a mean of about 39–40 hairs at which value it remained stable for the next three generations showing no further response to the selection pressure being continuously applied. But between generations five and seven there was a second and much larger response leading to a new 'plateau' at about 44 bristles. The initial response to a level rather higher than the higher of the two parental lines was probably due to reassortment of genes which were not linked, that is genes on different chromosomes, while the second and larger response probably followed a recombination between linked genes. This and many similar experiments show the large hidden potential variation which can exist even when the phenotypic variation is very small, in this case the two parental lines being very similar. While the starting material in this experiment consisted of two highly inbred laboratory strains other selection experiments have shown that lines can be selected from natural populations which go outside the original phenotypic range.

4.4   Local adaptation and migration

The 'release' of variation resulting in the creation of novel genotypic combinations can only happen if recombination occurs and recombination is a possibility only if multiple heterozygous individuals occur. And so we return to the problem of the maintenance of genetic heterogeneity at individual genes. Previously this was discussed in the context of a single population and some of the ways in which polymorphisms can be maintained were considered. Additionally the variation may occur spatially with one allele being predominant in one population and a different allele in another. If these two groups do not interbreed then they are genuine genetic populations. But this does not mean that they must always remain so, for a change in environmental circumstances could cause a breakdown of the reproductive isolation and produce a new unit of potential evolutionary change. The potential to store genetic variation in this way does not depend upon complete reproductive isolation of the two groups, it can also happen if the degree of migration between them is small. Of course storage occurs only if different alleles are common in the two groups and we may then ask how such divergences can arise.

One possible cause of such differences between populations (or if there remains a degree of migration between them they could be called sub-populations) is divergent selection, that is selection favouring one extreme of a character in one group, but the other extreme in the second group. That this process has happened in the recent past can be seen from investigations by Professor A. D. Bradshaw and his colleagues on the spoil tips of certain disused mines in North Wales. Until it became uneconomic to do so in the nineteenth century there was mining activity in various parts of North Wales for certain heavy metals, such as lead and copper, which are elements that are toxic to plants if present in soil in anything over a few parts per million. Inevitably the waste produced by these mines and deposited in the spoil tips was heavily contaminated, but since the mines stopped working the tips have been gradually, although still sparsely, recolonized by a number of species. Among these colonizers is the grass *Agrostis tenuis* which is frequently common in adjacent pastures and BRADSHAW (1965) has shown that plants collected from the mine tip are, in laboratory experiments, capable of tolerating much higher soil concentrations of heavy metals than are plants from adjacent uncontaminated pastures. Conversely the pasture plants grow more vigorously in normal soil than do the mine plants. Thus selection appears to be divergent and the interesting feature of this situation is that the two groups occur within a few metres of each other, for the transition from spoil tip to pasture is sharp. Further studies have shown that despite this proximity the two groups have much reduced migration between them for there is a difference in flowering time so that pollination is much more likely to occur within groups than between.

This study represents a case where there is an attempt to find a genetic difference which might be correlated with an obvious environmental discontinuity. In many other studies a spatial genetic difference has been the primary observation and this has led to an attempt to find an associated, and hopefully a causal, environmental variable. The most promising situations are those in which the genetic variation shows a gradual trend in one dimension of space. Melanic forms of *Biston betularia*, for example, are very frequent in the industrial area around Liverpool but the proportion in each population steadily declines as we move westward into Wales and it is very low at Bangor. Such a gradation is called a cline.

Many clines have been studied involving a wide variety of characters but recently there has been great interest in the search for clines of the electrophoretically detectable variants of an enzyme or protein. The catfish, *Catostomus clarki*, in North America shows variation in the electrophoretic properties of an esterase enzyme. Two different alleles of the gene producing the enzyme can be detected: one produces a fast migrating form of the enzyme (Es-i^a), the other a slow form (Es-I^b), while the heterozygote shows both forms (Fig. 4–3a).

The distribution of these genotypes throughout the Colorado river system (including its tributaries) has been surveyed by KOEHN (1969) and shows a north–south cline. In the more northerly populations the Es-I^b allele predominates while the southern populations are generally homozygous for the Es-I^a allele. Between these areas there is a cline of gene frequencies (Fig. 4–3b). It appears that in the southern population the Es-I^a allele is strongly favoured while in northern areas selection is in favour of the Es-I^b allele. In the intervening zone which shows the cline selection pressures may be very small and gene frequency in any particular population may depend solely on the relative migration received from the two homogeneous areas. Alternatively this effect of migration may be combined with a form of selection, such as heterozygote advantage, which can actually maintain a polymorphism. A latitudinal cline of this kind may well involve temperature as a selective force and there is some experimental evidence that this is so in the *Catostomus* case. Esterase enzymes from the three genotypes have been shown to differ in their optimum temperatures: the Es-I^a/Es-I^a genotype produces the most active enzyme at high temperatures, the Es-I^b/Es-I^b at low, while the enzyme from the Es-I^a/Es-I^b heterozygote is the most active at intermediate temperatures (Fig. 4–4). Not only is this in keeping with the direction of the cline, with the allele most common at the southern end of the range having the enzyme with the highest optimum temperature, but it also suggests that heterozygote advantage could occur at intermediate temperatures. While the detection of such correlations is a long way from providing proof of causation, it is suggestive.

But even in the absence of clear evidence of the effects of an environmental variable we may still be able to conclude that it is highly

32

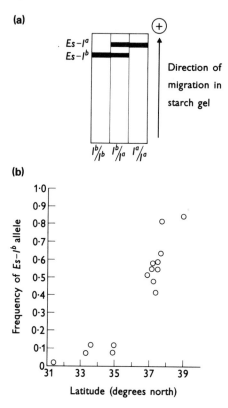

(a)

$Es-I^a$
$Es-I^b$

Direction of
migration in
starch gel

$I^b/_{I^b}$ $I^b/_{I^a}$ $I^a/_{I^a}$

(b)

Frequency of $Es-I^b$ allele

1·0
0·9
0·8
0·7
0·6
0·5
0·4
0·3
0·2
0·1
0

31   33   35   37   39
Latitude (degrees north)

Fig. 4–3 (a) Starch gel phenotypes of esterases in *Catostomus clarki*. (b) Distribution, against latitude, of $Es-I^b$ esterase allele. (Reproduced, with permission, from KOEHN and RASMUSSEN, 1967. Copyright, Plenum Press, London.)

likely that natural selection is responsible for a particular distribution of genetic variation from the pattern of that distribution. Nevertheless, if we do not know what factor is responsible it is unlikely, although not impossible, that we can have a clear idea of how it works on the genotypic differences. And a knowledge of whether or not, in the *Catostomus* example, there is heterozygote advantage along the cline is important in speculating on the likely evolutionary outcome. If heterozygote advantage does occur then the cline is likely to be stable unless there are considerable climatic changes, but if there is no heterozygote advantage and the cline is maintained by migration this may be unstable. In such a situation there would be an advantage in a reduction in the amount of migration which is tending to introduce unfavourable alleles into both populations. A reduction in the migration would be achieved by a decrease in the degree of interbreeding between the two populations, a change which seems to have happened to the *Agrostis tenuis* populations in adjacent mine and pasture habitats. The occurrence and evolution of reproductive isolation will be discussed in some detail in the next chapter.

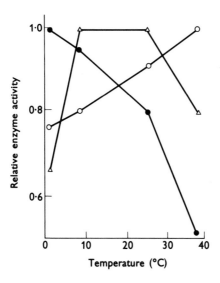

○ Enzyme extract from $Es-I^a/Es-I^a$ homozygotes
△ Enzyme extract from $Es-I^a/Es-I^b$ heterozygotes
● Enzyme extract from $Es-I^b/Es-I^b$ homozygotes

Fig. 4–4 Activities at different temperatures of esterase extracted from the 3 esterase genotypes of *Catostomus clarki*. (Reproduced, with permission, from KOEHN, 1969. Copyright 1969 by the American Association for the Advancement of Science.)

5 Formation of Races and Species

5.1 The concepts of race and species

In the previous discussion on natural selection and the maintenance of variation an attempt was made to develop that concept of population which is held by geneticists studying evolution. But it must also have been clear that, while this concept of an 'ideal' population is a very useful one both for theoretical treatments and as a basis for direct investigations, there are many situations where it is difficult to apply and may indeed be rather misleading to try to do so. Our ideal population is one in which the probability that any two individual members of that population will mate is not dependent upon the distance between them; these probabilities are either equal for all possible pairs (random mating), or if not equal are determined by the degree of genetic relationship (inbreeding) or of phenotypic similarity (assortative mating). In practice, the distance factor is important even in highly mobile organisms such as birds, many species of which show territoriality. There occurs a complete range of situations from a population in which mating is at random to one in which there are two or more populations between which there is only rare migration.

The rate of migration is crucial. If it is high the two groups will be genetically similar, while if it is low they may be genetically very diverse. Since it is not easy to determine the migration rate it is often difficult to decide whether the two populations should be treated as one or as two, and in fact the criterion used is often the degree of genetic difference between them. So we have two parameters for which we need quantitative estimates; the *rate* of migration and the *degree* of genetic diversity. The former can be defined as the proportion of genes introduced from one population to the other in each generation, while the latter can be expressed as differences in gene frequencies at a number of loci. But this latter point introduces another problem: which loci should be considered? Ideally one would know the frequencies of the different alleles at all loci of the organisms, but of course this is impossible in practice for there is no organism (except a few very small viruses of certain bacteria) in which all the genetic loci have even been identified, and even if all loci were known the practical task of determining the complete genotype for all of several thousand loci of all the individuals in a reasonably sized sample would be quite impossible. Thus we have to decide to study a limited number of known loci, and this choice will be a subjective one. It may be that a particular investigation is only interested in variation in a certain character and so the genes studied will be those which determine this. But the information obtained may be of doubtful value in attempting to decide whether two groups constitute a single population or two. For some loci there may be large differences in gene frequency between the populations, and they may even be 'fixed' for different alleles, while other loci show great similarity. In the latter case the similarity will occur in one of two different ways: both populations may be polymorphic with similar gene frequencies, or both may be monomorphic for the same allele. We shall return to this point later when considering some examples.

This rather abstract discussion has so far begged the important question of the criteria used in the first place to distinguish these two groups, about which we have been concerned as to whether they constitute a single genetic population or two. Behind such an observation there will usually be a mixture of ecological and genetic factors, the ecological criteria being that the two groups occupy distinct areas (which may be geographically separate) between which there is variation in one or more environmental elements, while the genetic factor is that the two groups show certain phenotypic dissimilarities that are known, from breeding tests, to have a genetic basis.

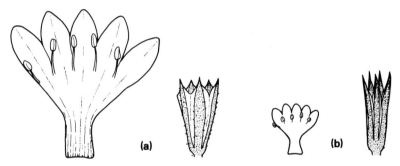

Fig. 5–1   Flowers of the sun race (achilleaefolia) and the shade race (multicaulis) of *Gilia achilleaefolia*. (Reproduced, with permission, from GRANT, 1963 and the publisher.)

When two such groups of a single species show considerable genetic differences and where they occupy ecologically very distinct or geographically well separated habitats they may be called races. The nature of the habitat distinction leads to the concepts of ecological and geographical races. Of course it is likely that the regions occupied by two geographical races will also be characterized by a number of differences in environmental factors and so we could say that geographical races are ecological races which are spatially clearly separated.

An excellent example of ecological races within a species has been provided by the studies of GRANT (1963) on the plant *Gilia achilleaefolia* in the Californian Coast Range. Two races are known in this area: a sun race and a shade race, with a morphological difference in flower structure between them although intermediates also occur (Fig. 5–1). The map in Fig. 5–2 shows the distribution of the two races; individual populations being classified on the morphological criteria. The interspersal of the races reflects the heterogeneity of habitats within this region, for the sun race (achilleaefolia) occurs on open grassy slopes while the shade race (multicaulis) is found in shady oak woods or thin redwood groves. Thus the two types occupy ecologically distinct environments. That the morphological difference is genetically based is shown by crosses between them.

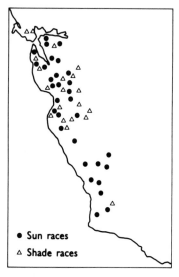

Fig. 5–2 The distribution in California of sun and shade races of *Gilia achilleaefolia*. (Reproduced, with permission, from GRANT, 1963 and the publisher.)

Many examples of geographical races could be given: the races of man are essentially geographical. There is a great deal of information about differences between these races which can be used to illustrate some general features of races and their formation. The characters which are used for description of human races can be divided into two groups. The first includes the obvious differences in skin pigmentation or morphology, this group of characters having a complicated genetic basis. Characters in the second group have simpler inheritance, often due to the segregation of a single gene, and include many 'biochemical' differences such as blood groups, or the ability to taste phenyl-thio-carbamide. The existing classification of races has been based on the first group and the characters which have been given most weight are those which tend to emphasize differences between races. The choice of characters can be of crucial importance to the kind of classification made.

Fig. 5–3 shows the situation revealed by attempts to quantify one of the characters most commonly used in racial classification, skin pigmentation. These curves are the distribution of relative darkness, as measured by degree of reflectance of light, in two groups, American Whites and American Negroes. It is clear that not only is there variation within each group but that there is an overlap extending across a considerable range. One of the reasons for this is that intermarriage

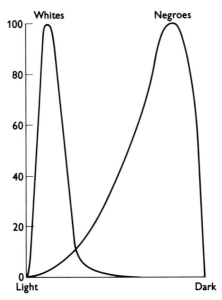

Fig. 5–3 Frequency distribution of skin pigmentation, as measured by reflectance of light, in samples of American whites and American negroes.

between races in the U.S.A. has occurred and the distinctions between them are breaking down. That this is so can be seen more clearly by comparing these groups for a number of different individual genes. Estimates of gene frequencies within both American Whites and American Negroes have been made for a number of loci and in some cases it is possible to compare these with the frequencies now occurring in West African Negroes from whence the American Negroes came. Table 2 shows that in general American Negroes have frequencies intermediate between those of the other two groups. If we assume that the West African populations have changed little since the seventeenth and eighteenth centuries and that the frequencies in the white population have been little affected by intermarriage (a reasonable assumption since whites outnumber negroes in the U.S.A. by about 8 : 1) there are two factors which might account for the American Negro frequencies. These are selection and migration of genes from the white population. Selection in a new environment may operate on both populations to tend to produce a convergence of gene frequencies, but an important point is that it is

Table 2 Comparison of gene frequencies in West African Negro, American Negro and American White populations

Gene	Allele	West African Negro	Frequencies American Negro	American White
Rh	R^0	0.594	0.535	0.037
	R^1	0.069	0.103	0.426
	R^2	0.086	0.108	0.148
	r	0.211	0.230	0.358
ABO	A	0.148	0.158	0.246
	B	0.151	0.129	0.050
	O	0.704	0.713	0.704

unlikely to be of equal strength on all loci. That there occurs a group of loci with a similar 'movement' from West African towards white gene frequencies suggests that they are all changing through the influence of a single factor; and intermarriage will have the same effect on all differences in gene frequency.

Thus we can from this example point to an important principle: that races are not permanently separated groups. If two races, which had previously been reproductively isolated through certain geographical or ecological factors, come into contact again because of a change in environmental conditions, they *may* interbreed. If they do, any genetic differences will tend to disappear, but the interesting questions are *whether* they will interbreed, and *what factors* will determine this? Before turning to a discussion of reproductive barriers to interbreeding let us consider a few 'case' histories of geographical races to illustrate the range of phenomena with which we are concerned.

5.2   Geographical differences

The nature of the geographical distribution of plants and animals was one of the main props in Darwin's arguments for his theory of evolution, for he contended that the theory could make sense of the existing distribution. In particular he was greatly impressed by the diversity in oceanic islands and his studies of the fauna of the Galapagos Islands now provide classic examples of divergence following colonization. Fundamentally the argument is that following a rare (and possibly unique) colonization from the mainland by a very small number of a particular type of organism, adaptive radiation occurred to fill a wide variety of ecological niches which on the mainland would have been occupied by other species. Thus amongst Darwin's finches in the Galapagos Islands there exist tree dwellers and ground dwellers, insectivores and herbivores. A more recently studied and in some ways even more remarkable phenomenon is the enormous diversity of species of *Drosophila* which exists on the Hawaiian Islands. At the latest count (the number continues to increase as more species are discovered) there are about 400 species of Drosophila (and of a closely related genus) in Hawaii. This, in fact, accounts for about one third of *all* known species of this group. Furthermore, while in some features, especially external morphological characters, the Hawaiian species show great diversity, in other ways (internal structure and chromosome characters) they are remarkably alike; so much so that the entire range of Hawaiian Drosophila might have evolved from a *single* colonization (perhaps from Japan) occurring not more than five million years ago. This is the earliest possible date for the colonization because the islands are volcanic and the oldest was formed at that time. These island diversifications were possible because the colonizing populations were isolated from mainland populations and became adapted to fill a great variety of ecological niches which were occupied in the continent of origin by other dipteran insects. But clearly this is not a sufficient condition, for the question arises: how was it possible for so many species to arise from a common ancestor in such close proximity? And why indeed do we describe these different Hawaiian Drosophila as species rather than races? If in some respects (chromosomes and internal structure) they are so alike, but in certain external features they are so diverse, by what criteria do we conclude that the morphological features should be given so much more weight? The answer is that, although in taxonomy external morphological features are commonly used, the modern neo-Darwinian concept of species does not rely on any particular phenotypic attribute but on the degree of reproductive isolation which occurs between two groups.

5.3   Barriers to breeding between groups

In man the various races are not reproductively isolated. We consider that the races have evolved in geographical isolation from each other but that when they have been in contact interbreeding has occurred. In recent centuries this has been an increasing phenomenon because of much improved transport. But in other organisms two groups which are so alike that it seems they must have had a recent common ancestor may not inter-breed. *Drosophila pseudoobscura* and *D. persimilis* are so alike morphologically that only an expert can classify them, and their geographical distributions overlap in the western U.S.A., yet they do not inter-breed. What is the barrier to inter-breeding?

When considering the mechanisms which lead to reproductive isolation we need to remember that the important criterion for isolation is that there shall be no exchange of genes between the groups; each one maintains its genetic integrity. Thus a reproductive barrier can be said to occur even if hybrids between the two groups are formed, provided that these hybrids do not themselves breed, either because they do not survive to reproductive maturity or because they are sterile. Clearly reproductive isolation exists if no hybrid zygotes are formed because fertilization cannot occur, and there are a number of mechanisms which will prevent fertilization. Thus we can produce a classification of systems which act as barriers to inter-breeding:

> A. Geographical isolation
> B. Reproductive isolation
> > 1. Prezygotic barriers
> > > (i)   Seasonal
> > > (ii) Mechanical
> > > (iii) Ethological
> > > (iv) Incompatibility
> > 2. Postzygotic barriers
> > > (v)   Hybrid inviability
> > > (vi) Hybrid sterility

Let us now briefly discuss these six types of mechanisms:

(i) Seasonal barriers. This is a fairly obvious mechanism which will operate, for example, where two species of plants flower at different times of the year. *Pinus radiata* and *P. attenuata* have generally different distributions but they do overlap at Monterey Bay in California. Although they can be artificially crossed to produce hybrids, crossing in nature in Monterey is very rare because in *P. radiata* pollination occurs in early February, while in *P. attenuata* this does not happen until April.

(ii) Mechanical isolation. There are a number of cases in plants in which the flowers are adapted to different pollinators but where artificially produced crosses would be fertile. *Aquilegia longissima* has erect pale yellow flowers with long thin spurs and is pollinated by hawk moths, while *A. formosa* has pendant red flowers with short stout spurs and is pollinated by humming-birds. Hybridization does not occur in nature although these two species are inter-fertile. Mechanical isolation occurs in animals where the genitalia differ to an extent that prevents effective copulation.

(iii) Ethological isolation. This is of obvious importance in animals where courtship displays occur. In the genus Drosophila for example males show species-specific pre-mating activities in order to persuade females to mate. A single *D. melanogaster* female given, in an experimental situation, a choice of two males, one of which is from the same species while the other is from the species *D. persimilis*, will regularly mate with the melanogaster male.

(iv) Incompatibility. Even if two species are reproductively mature at the same time and if there are no mechanical or ethological barriers to the transfer of gametes between them there may still be a final defence against the formation of hybrid zygotes. This will depend upon some kind of physiological incompatibility between the gametes of the two species or,

as in angiosperms, between male gametes of one species and maternal tissue of the other.

(v) Hybrid inviability and *(vi) Hybrid sterility.* If the hybrid zygotes do not themselves breed, then there will be no 'gene flow' between the two species which will then remain genetically isolated. Hybrids may not breed either because they do not survive until reproductive maturity or because they are sterile. A classic case of a sterile interspecific hybrid is the mule, the product of a cross between a horse and a donkey.

This very brief discussion of the mechanisms leading to reproductive isolation between species can provide no more than a framework, and fuller treatments can be found elsewhere (GRANT, 1963; DOBZHANSKY, 1970). What is more important in this context is to consider how the neo-Darwinian theory of evolution can account for the existence of these barriers.

There are three alternative processes which could occur. Firstly, two groups of organisms may become geographically separated and in the course of differential adaptation it may come about that the changes are such as constitute a reproductive barrier. For example, two groups of plants which became adapted to different locally available insect pollinators would probably remain genetically isolated from each other if the geographical separation broke down.

Secondly, the initial geographical differentiation may not proceed to the point where crossing is impossible but may lead to a reduction in crossing potential. If then these two populations coincide in the same area but if hybrids between them are not well adapted to either of the ecological niches occupied by the parental groups, selection may operate to continue the trend towards reproductive isolation and the development of more efficient barriers. This reinforcement process is often called the 'Wallace effect' after A. R. Wallace who first postulated it.

Thirdly, a single population with the possibilities of adapting to two different ecological niches within the same area might split into two separate species with the evolution of a reproductive barrier between them.

Speciation arising by the first two processes is said to be *allopatric*, and by the third process, which at no stage involved geographical separation, is called *sympatric*. The first process would depend entirely on chance, the second would depend upon chance initiation but development by selection, while the third mechanism would be entirely determined by selection. It should be emphasized here that the use of the term 'chance' is meant to imply only that the reproductive isolation is a chance side effect of other changes, not that these primary changes have occurred by accident. Indeed it seems most likely that the primary differentiation between the two groups will have been produced by selective forces.

There has been much discussion amongst evolutionists of the question: which of these three processes has been the most common in nature? The evidence used to attempt to answer this question has been obtained partly by observation of natural situations which would *appear* to be on the way to speciation (although there can be no guarantee that they might not revert in future), and partly by experimental attempts to demonstrate the evolution of reproductive barriers. The need to keep this discussion concise precludes the use of a wide range of examples and only two will be given.

In the plant genus *Phlox*, the common flower colour in *P. glaberrima* is red as it is in populations of *P. pilosa* which are allopatric. But *P. pilosa* populations which are sympatric to *P. glaberrima* are sometimes white and such populations show little hybridization between species.

The second example concerns an experimental demonstration that reproductive isolation can arise sympatrically. PATERNIANI (1969) began an experiment with two varieties of maize, each variety carrying a different recessive mutation. Thus the two lines were recognizably distinct from each other, as also they were from the hybrid between them which would have the dominant phenotype for both characters. When interplanted in a field experiment the two lines showed about 50% intercrossing, that is about half the seed produced by each line had the double dominant phenotype and must be of hybrid origin. Paterniani selected those plants within each line which had the lowest proportion of hybrid seed and used these selections as the parents for the next generation. In this next generation he repeated the process, and continued this for six generations. After six generations of selection the two lines were now intercrossing with each other to the extent of only about 5%. Thus they were rapidly becoming reproductively isolated and the barriers which had arisen were, firstly, that the flowering times now differed by about ten days, compared to almost perfect synchrony at the beginning, and, secondly, that the pollen of one of the lines was now much less compatible with the stigma of the other.

The experimental demonstration that reproductive barriers can arise easily indicates that this *could* happen in nature, but of course it does not give us any information on *how often* speciation may have occurred sympatrically. Indeed it is very difficult to attempt to reconstruct the most likely sequence of events to have led to the existence of any given reproductive barrier between two species. This is a general problem which besets students of evolution and it has led some scientific purists to argue that the study of biological evolution cannot be a truly scientific enterprise at all because it represents an enormous number of unique events. But to take this extreme viewpoint is to fail to recognize the wood for the trees because, while the detail will be unique in each case, the modern theory of evolution has shown that general principles can be extracted, such as natural selection, and that a theoretical framework can be constructed upon these principles. Nevertheless, the problem of applying these principles to particular cases is extremely difficult. For example, where a genetic difference exists between two populations has this been produced by natural selection, by random genetic drift, or by some combination of the two? Sometimes, such as when changes are rapid enough to be detected, it is possible to fit a theoretical model to the observations and to show that there is a regularity about the phenomenon, as in the case of industrial melanism, which can be explained by selection but not by random changes. In addition it may be possible to demonstrate experimentally that a certain environmental factor has selective consequences; again this has been done for industrial melanism.

Returning to the topic of speciation can we say how often species barriers have arisen as a result of the action of each of the three alternative processes we have outlined, and which we could call: 'chance', 'chance/selection' and 'selection'? Associated with the name of MAYR (1942) has been the commonly held view that all, or very nearly all, speciation was due to the first or second process, that is it was essentially allopatric. More recently, largely because of experiments like that of Paterniani, it has begun to be accepted that sympatric speciation may have been a commoner process than was once thought, while still allowing that allopatric processes would be the most frequent. It is at present meaningless to attempt to be more quantitative than this very general statement. What is not in doubt is that natural selection has been a universal force in developing reproductive barriers at least to reinforce a degree of isolation which may initially have arisen by chance.

6   Some Problems to be Answered

Up to this point in the book the concern has been to present the main points of the modern form of Darwinism. In doing this an attempt has been made to stress three threads which are interwoven to form the main fabric of the theory. These are: (i) the nature of the underlying genetic variation, a nature which in the past two decades we have come to understand much better because of the findings of molecular biology; (ii) the ways in which reproductive systems organize this variation into genotypes of individuals and genetic compositions of populations; and (iii) the operations of natural selection producing genetic differences over time and space and in many cases leading to the creation of races and species. These three topics are interwoven in that natural selection acts on the creation (through mutation rates) and organization (through reproductive systems) of genetic variety, while the state of this variety in turn conditions responses to natural selection.

Most of the content of these first five chapters is uncontroversial and it is likely that most professional students of evolution and genetics would agree with the main outlines of the framework, although there would be vigorous dispute over many details. But although there might be general agreement over the acceptability of the theory to explain the phenomena which have so far been discussed, many large problems remain for which the applicability of the theory is uncertain or controversial.

6.1   Are evolutionary changes sometimes at random?

This topic has already been introduced in relation to the existence of polymorphic populations when the form in which the question was put represented the other side of the coin, namely 'are all differences adaptive'? Under that heading the problem was considered of whether or not an observed polymorphism was maintained by natural selection or had arisen through chance factors which had nothing to do with adaptation. If the allelic variation is indeed neutral with respect to fitness the allelic frequencies will be determined by the chance effects of limited population size. In two separate populations different alleles may become fixed by such chance processes.

There is little doubt that such events might happen or that on occasions they have happened; but there is considerable controversy over how frequently specific differences might have arisen by chance. The problem is this: how to establish that for any particular characteristic the difference between any two species has occurred because of random non-adaptive changes. The argument that some evolutionary changes may have happened in this way has recently developed considerable force because of information from molecular biology. In fact it has been claimed that the evidence from molecular biology shows that such events may have been sufficiently common that we ought to recognize this kind of evolution as an important and distinct category by naming it 'non-Darwinian' evolution!

Much of the discussion has centred on the amino acid sequences of certain proteins. Cytochrome c is a protein in the electron transport chain associated with respiration and is found in all organisms from yeast to man. Although the number of amino acid links in the chain varies between 104 and 112 there are about 30 positions which are always filled by the same amino acid. The remaining 70 or so positions vary between species and, not surprisingly, there are fewer differences when comparing closely related species than for distant relatives. But this association turns out to be quantitatively even more interesting (DICKERSON, 1972). Table 3 compares man with six widely divergent species, giving paleontological

Table 3   Comparisons of cytochrome c sequences between man and other species

Species	Estimated time since common ancestor (million years)	Number of variant amino acids
Rabbit	75	9
Pigeon	300	12
Bullfrog	350	18
Dogfish	400	24
Drosophila	approx 600	28
Wheat	more than 1000	40

estimates of the time of divergence of the two evolutionary lines in each comparison, together with the number of amino acid differences in the cytochrome c. There is clearly a correlation between the number of amino acid differences and the time since divergence, and if, by considering the genetic code, these differences are converted into the probable number of genetic changes the correlation becomes a strongly linear one. The basis for such a correction is that, while in a comparison of two recently diverged species a given amino acid difference is likely to represent a single mutational event, a single difference between two long separated lines may be the end product of several mutations. In some cases the genetic code tells us that a difference must have involved more than one mutation. Amino acid position 22 in cytochrome c is occupied by lysine in man, asparagine in dogfish and alanine in wheat. The RNA codons for these three amino acids are:

man:	lysine	AAA or AAG
dogfish:	asparagine	AAU or AAC
wheat:	alanine	GCU, GCC, GCA or GCG

Thus the difference between man and dogfish could have arisen as the result of a single mutation in the third base of the codon, but the vertebrate-plant difference must have involved at least two mutations in the first and second bases.

When allowance is made for such effects and the sequence of cytochrome c (and other proteins) are compared in terms of the minimum number of mutations, the correlation with time is surprisingly linear for any one protein. Such correlations suggest that the rate of change of amino acids at any one of the 70 or so variant positions is linearly related to time and the 'non-Darwinian' evolutionists argue that this is just what is to be expected if the substitutions at these positions are due to neutral mutation followed by random genetic drift. Selection on the other hand is much more likely to occur irregularly with a series of rapid changes, say when there is a major environmental change or when a new phylum is established, interspered with periods of little change.

Interpretation of these observations is very controversial with those who doubt that such random evolution happens frequently arguing that the correlation is misleading for two reasons. Firstly, because it is inexact, for the time span estimates are very imprecise; and secondly, because it obscures the real story, for a detailed analysis shows that many of the sequence substitutions involve amino acids which are chemically much more similar than would be expected if changes were at random. This implies, it is argued, that selective constrains have been operating at the variant sites. It is generally agreed that the constancy at the thirty invariant positions in the protein imply strong selection forces acting to maintain the functional properties of the cytochrome c.

6.2   Is natural selection sufficient?

The argument about the extent of random changes is but one of a number of sources of criticism of the neo-Darwinian view which gives a central role to natural selection. The 'non-Darwinian' viewpoint is that natural selection could not have produced the enormous range of living organisms in the time available since life first appeared and that random changes fill this gap.

Another attack on neo-Darwinism comes from the opposite flank. This is the view that natural selection acting on variation produced by random mutation could not have produced the range of beautifully adapted organisms which are found in the living world. In extreme forms this becomes subjective and mystical and there are such elements in Lamarckism which conceives of organisms 'striving' to become better adapted to the environment. One component of Lamarck's theory was that evolution would proceed through the inheritance of acquired characters and in the first chapter one of the most famous experiments which has been held to support this theory was discussed. This experiment by Kammerer showed that, after several generations of being forced in the laboratory to breed in water, a stock of the Midwife Toad (*Alytes obstetricians*) produced males with nuptial pads which are normally found only in related species which are naturally water breeding. But because there was no guarantee that the initial stock was genetically homogeneous it remains very likely that selective forces produced by the treatment allowed only those males with some nuptial pad development to mate successfully.

In an earlier discussion on selection it was shown how laboratory experiments have demonstrated that very large responses may be obtained from what appears to be a phenotypically rather homogeneous stock, phenotypic similarities overlying marked genetic differences. There is one particular kind of laboratory selection experiment which is particularly relevant in this context for it shows how an apparent case of inheritance of an acquired character is really due to selection. One example concerns the vein patterns on the wings of *Drosophila melanogaster*. Fig. 6–1 shows both a normal pattern and the disruption in a posterior

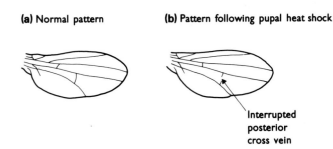

(a) Normal pattern   **(b) Pattern following pupal heat shock**

Interrupted
posterior
cross vein

Fig. 6–1   Wing vein patterns of *Drosophila melanogaster* in (a) fly reared at 25°C; and (b) fly from pupa which was given a 40°C heat shock.

cross vein which occurs in certain individuals if the pupal stage is given a heat shock. In a particular laboratory stock WADDINGTON (1953) found that 34% of the flies showed this reaction to a heat shock, the remainder being unaffected. He selected his parents from among those affected and repeated the treatment on the next generation. This process was repeated over a number of generations until at the 24th generation the proportion of flies responding to the heat shock had risen to 94%. Furthermore he found that after this selection to increase the proportion responding, the new stock actually contained some flies which had broken cross veins even without the heat shock. That this was due to selection on genetic variability was shown by a control line breeding separately from those flies which did not respond. Here the incidence of flies affected was reduced to 17% and this selected stock contained no flies which had broken cross veins when reared at a normal temperature. A detailed study of the two lines showed that they did differ genetically.

These results can be explained if the presence of a complete cross vein is a threshold character. At normal temperatures all flies in the base stock were above the critical level, but a heat shock so raised the threshold that 34% of the stock had genotypes which put them below the minimum level. Selection amongst these produced a stock which not only had more flies below the high threshold due to the heat shock but also a few which were below the lower threshold level occurring at normal temperatures (see Fig. 6-2).

This type of phenomenon, which is superficially like the inheritance of acquired characters but can be readily explained in terms of selection of Mendelian genes, has been called 'genetic assimilation'.

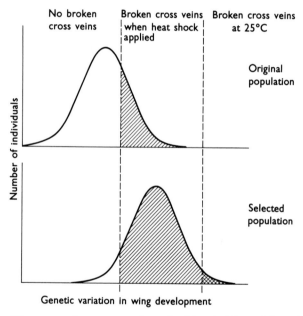

Fig. 6-2 Diagrammatic representation of a 'threshold' model to show how selection for a higher incidence of flies which respond to a heat shock might produce a few which produce broken cross veins even without a heat treatment.

6.3 How do new functions arise?

Experiments in genetic assimilation have shown that the responses can only occur if the initial stock is genetically heterogeneous. In the wing vein studies control experiments were set up using a highly inbred laboratory stock, which would be expected to have very little genetic variation; selection on this stock failed to produce a response. Thus the responses depend on pre-existing genetic variation which was not induced by the environmental factor. But how can we account for the occurrence of this variation in the base stocks? The breakage of a posterior wing vein would seem to be the loss of a function and so the available variation could be produced by rare but recurrent mutations in the genes which control the vein formation. This pushes the question back a stage further: how did these genes evolve their function of producing a vein in the first place? The problem of the origin of particular functions is perhaps at its most obvious with such highly adapted organs as the vertebrate eye. To account for this according to neo-Darwinism we have to postulate that there was a step by step evolution by natural selection acting on variation produced by random mutation, at each step there being a slight but positive advantage over the existing form. That this should have happened may seem intuitively implausible, but is clearly less so than that the eye evolved in a single change.

At a more fundamental level we may ask how the genes determining a particular function or structure arose. There are two possible explanations: one is that the gene had been performing some other function but through a series of mutations and natural selection it lost its original function and gained a new one; the alternative is that it was a genuinely new gene, that is a new piece of DNA. The problem with the first explanation is that there would most probably be an intermediate stage when it had lost its original function but not yet gained the new property; such mutant alleles would be inferior and be eliminated by natural selection. The problem with the second view is what mechanism could produce the extra DNA in a form which is likely to have any biological activity at all? This dilemma could well be resolved by the known tendency of the DNA to produce 'duplications'. If a particular region of the DNA constituting a gene were to become duplicated, the organism would have an extra and redundant copy of that gene. Mutations could now accumulate in one of these copies without the constraint imposed by natural selection acting to maintain the original function, which would be carried out by the other copy.

A limited number of experiments to test this hypothesis have been conducted in bacteria by forcing growth on a medium containing a chemical, such as the sole carbon source, which is a novelty to the bacterium, that is one which has not been previously encountered by the organism. *Klebsiella aerogenes* can grow on a medium containing the sugar ribitol as a sole carbon source but the wild type can grow only very poorly on the related sugar xylitol, and the enzyme which acts on ribitol has only low activity on xylitol. Using a continuous culture technique RIGBY, BURLEIGH and HARTLEY (1974) found that after some time with xylitol as the only carbon source the wild type strain could grow faster on it. Analysis of this strain showed that there was now much more of the enzyme although its activity against xylitol (expressed per unit weight of enzyme) had not changed. In other words the newly evolved strain could grow faster because it had more, rather than a better, enzyme. The reason it had more enzyme was partly because it had two, or in some cases even three, copies of the gene for the enzyme. This experiment has demonstrated the occurrence of the first part of the mechanism postulated in the duplication hypothesis; the duplication itself has arisen but random changes in one or other copy have not yet accumulated to a level which will produce an improved activity against xylitol.

6.4   The origin of life

The creation of a new function, such as a novel enzyme, represents a discontinuity in evolution, which a satisfactory theory should be able to explain. An attempt has been made to outline some current ideas on how new enzymes might arise, which could serve as a model for the origin of other new properties and functions. But how can we explain the largest discontinuity of all, the origin of life itself?

As with so many evolutionary topics we find that molecular biology has made considerable contributions to an understanding of the probable sequence of events leading to the first living organisms. It is not surprising that this is so because we are concerned with essentially molecular phenomena. Although there are many gaps in the model which most likely represents a reconstruction of the origin of life, the story which is being created can be briefly summarized as follows.

The age of the earth is about 4500 million (4.5×10^9) years while the oldest known rocks are 3.0×10^9. In these rocks have been found traces of organisms which are similar to present day blue-green algae so that by the time the earth was 1.5×10^9 years old cellular organisms existed. What were conditions like in this period? It is now generally accepted that the atmosphere would have been quite unlike its present form and would have lacked oxygen, but be composed of nitrogen, methane, carbon dioxide, water vapour and ammonia. Thus it would have been chemically a 'reducing' atmosphere. One of the lines of evidence in favour of this view is the presence in the oldest known rocks of iron in the ferrous or reduced form whereas more recent rocks tend to contain much ferric (oxidized) iron. The transition to an atmosphere containing so much oxygen almost certainly had to wait upon the evolution of photosynthesizing organisms which could hydrolyse water and liberate oxygen. The importance of the existence of a reducing atmosphere is that laboratory experiments have shown that it is relatively easy to generate organic molecules in such an atmosphere but very difficult to do so in an oxidizing atmosphere. In these experiments electric discharges (simulating lightning) or strong UV irradiation (simulating solar irradiation before the evolution of an ozone layer which attenuates these wavelengths) applied to a mixture of the gases listed earlier created a number of organic molecules including amino acids and nitrogenous bases which are the basic building blocks of proteins and polynucleotides (nucleic acids) respectively.

The next stage must be the formation of the polymers (long chains of repeating units) and here it is not nearly as clear how this might have happened. In the laboratory it has not been possible to create spontaneous polynucleotides except by raising the temperature to well over 100°C. In living organisms the synthesis of such polynucelotides depends upon the catalytic action of specific enzymes. In our primeval environment these enzymes would not have existed but other inorganic catalysts might have. The organic molecules such as the nitrogeneous bases would have occurred in aqueous solution and through evaporation might have been locally concentrated. Another way in which they could have been concentrated is through absorption onto electrically charged mineral particles and it seems quite likely that some of these minerals could have catalysed the formation of polymers. Indeed there is the exciting, but highly speculative, possibility that some minerals such as certain mica clays might also have been involved in the next and crucial step, the evolution of self replication. In the formation of some of these clays the pattern of the distribution of molecules in an existing particle may determine the pattern in the newly formed layer by a kind of pattern printing process. If polynucleotides had formed on such particles they too might be replicated along with the clay and this could have been the beginning of a self replicating organic system.

But even if this idea is correct we are still a long way from achieving an autonomous living cell capable of catalysing its own replication by making its own enzymes. The belief that such changes were produced by natural selection may seem to be an act of faith rather than of science at present, but evolutionists feel confident that eventually the evidence will be available to support the general theory.

7 Can Evolution be Controlled?

The final question to be discussed concerns the possibilities of using our information and theories of how evolution proceeds to control future evolutionary events. The answer to the question 'Can evolution be controlled?' is 'Yes, in certain circumstances.' The achievements of such control are most obvious in those species, such as cultivated crops and domesticated animals, which man has directly utilized for his own benefit and convenience; the possibilities are present in many other species; while the attempts are most uncertain and most controversial in man himself.

7.1 Plant and animal breeding

Certain plants and animals have been used by man for thousands of years; wheat and barley are known from archaeological evidence to have been in cultivation in what is now Iraq as early as 5000 B.C. The selective forces operating on such species will have been modified in two ways. Firstly there will have been unconscious effects such as protection from natural predators which will have relaxed selection pressures towards evasion mechanisms; secondly, conscious selection will have occurred towards such aims as increased grain yield in wheat or faster hunters among dogs. Over thousands of years these processes have produced large changes from the original forms and enormous diversity among contemporary forms, as can be readily seen among dogs. Indeed it was among this great variation within plants and animals under domestication that Darwin found so much stimulus for the development of his ideas on evolution DARWIN (1899).

In the last few centuries breeders have used a more intensive and directional approach and in the last few decades a more scientifically based one. Despite these changes the effects of natural selection can never been in cultivation in what is now Iraq as early as 5000 B.C. The selective controlled experimental breeding programme natural selection will still operate and may indeed run counter to the imposed aims. Thus the breeding of plants, such as lettuce or grass, for production of vegetative parts may be in conflict with natural selection which will tend to divert the plants resources towards reproductive development.

The aims of a breeding programme are twofold. Firstly, it is hoped to increase by direct improvement the yield of whichever attribute is commercially (or aesthetically) important. Secondly, the features of the organism may be changed to allow it to escape from the effects of some restricting factors. The former aim is an obvious one; the latter may involve a number of different aspects. A common and serious disease can clearly be an important restricting factor and many breeding programmes would attempt to breed in resistance. A rather subtler demonstration of this second aim can be seen in one aspect of the well known 'Green Revolution'. Many crops will grow larger and more vigorously if fertilizers are added, and especially in response to the application of a suitable form of nitrogen. But the potential of this response cannot be realized in many varieties of such cereals as wheat and rice because the extra nitrogen causes them to grow so tall that the stems are unable to support the ears. The result is that the whole plant may fall over (lodge) before harvest and much of the grain is lost. The new varieties of rice and wheat produced by the Green Revolution breeding activities incorporated dwarfing genes which produce shorter plants which can benefit from extra fertilizer without lodging.

The practice of artificial breeding is essentially one of artificial selection, but selection will only produce a response if genetic variation exists. In most cases there will be genetic variation within the existing range of breeds or varieties, although if the character is one which is highly sensitive to environmental differences it may require sophisticated techniques to disentangle the genetic differences. Additional problems arise if the desired genetic variant is not available within existing cultivated or domesticated types. A search is then made among wild forms of that species or in related species. In plants some of the most productive areas for the occurrence of useful variation are those regions where the cultivated forms are the most primitive and may indeed show occasional crossing with the wild progenitors. The Russian botanist Vavilov recognized in the early decades of this century that there was a strong correlation for a particular crop between the areas where the greatest degree of genetic variation could be found (a centre of diversity) and the area where the crop was first taken into cultivation from the wild (a centre of origin). Fig. 7–1 shows the six major regions of domestication identified by Vavilov. Regions 4 and 5, the eastern Mediterranean and Southwest Asia (the Fertile Crescent) are known to be the regions in which wheat and barley were first cultivated and innumerable collecting trips to these regions have provided much useful material. One, but not the sole, reason for the great range of genetic variation in such regions is the existence of so many local cultivated varieties. That these occur is a consequence of the undeveloped local agriculture with individual farmers keeping part of the seed at harvest for next year's crop. As the agriculture becomes developed these varieties will largely be replaced by a few highly bred and genetically homeogeneous types and much of the storehouse will be lost. It is a paradox that the success of a 'green revolution' leads to the erosion of the base for future similar advances. Clearly there is a conservation problem here.

54

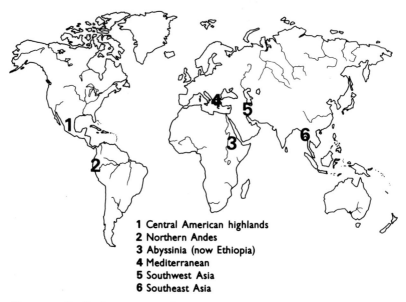

1 Central American highlands
2 Northern Andes
3 Abyssinia (now Ethiopia)
4 Mediterranean
5 Southwest Asia
6 Southeast Asia

Fig. 7–1 Vavilov's centres of plant domestication. (Reproduced, with permission, from BAKER, 1964.)

7.2 Conservation and evolution

The popular idea of conservation is of a concern with the continued survival of a rare species, and particularly if this should be a bird, a butterfly or an attractive flowering plant. The motivation for much of this concern is a mixture of academic, aesthetic and even moral considerations, but we have seen in the previous section that there are good practical reasons for trying to ensure the survival of at least certain parts of the range of organic forms which evolution has produced. And not all these practical requirements will be apparent now, for not only will new and presently unforeseen properties be demanded of existing cultivated species, but also the needs may be so novel as to require the cultivation of a hitherto wild or little cultivated species, as in the search for sources of unsaturated fats for human consumption.

That there is a need to take action to preserve the enormous existing range of genetic variation present in primitive cultivated races and in related wild populations and species is now widely accepted and is acknowledged by such organizations as FAO and the International Biological Programme. The problems which arise, apart from finding the necessary financial support, are to decide on the best strategies for conservation. Clearly, since the existing habitats will be destroyed the preservation must be done by cultivation in collections and botanic gardens. During such cultivation the combination of natural selection and genetic drift may erode some of the genetic variation. To minimize this the material should be grown at a number of diverse centres (to produce a range of divergent selective effects) and in large numbers (to reduce the effects of genetic drift).

In practice, decisions about the optimum strategy for conservation have to consider not only the optimum conditions for maintaining variation but also the available resources and the cost of the programme. Rarely will a collection be kept under conditions that are ideal 'genetically'; some compromise will have to be reached and a suitable slogan might be 'any variation is better than none at all'. A similar approach would be appropriate to the attempt to preserve in zoos or botanic gardens genuinely wild species which are threatened in their natural habitats. Here the criterion would be preventing the species becoming completely extinct without worrying too much about the representation of the complete range of variation found within the species. But even so there is a genetic component to survival chances for if the populations in captivity are small there must inevitably be a degree of inbreeding. And one of the common consequences of inbreeding a normally outbreeding species such as mammals is to reduce fertility, sometimes to a point where the population dies out. There is no answer to this problem except to keep the population size up in the hundreds or thousands, and clearly this is prohibitively expensive for large mammals in zoos. Game parks in their native regions where animals can be protected and can live as part of the natural ecosystem would seem to be the best solution.

7.3   Eugenics and the evolution of man

One of the most fascinating questions which can be asked of current evolution is to determine what changes are happening to man, and then to ask whether these changes can be modified by accelerating those which are favourable and reducing or reversing those which are deleterious. One form in which this question is framed is based on the widely held belief that the genetic stock of man is declining because we have so modified our environment that many stresses have been removed and because modern medicine now leads to the survival of some types which in the past would have perished.

Whatever the truth of such a claim, which shall be discussed later, there is no doubt that there are large numbers of heritable defects which the human species would be happier without. There is for example a condition called phenylketonurea which leads to severe mental retardation. It has been estimated that in European populations something like 1 in 10 000 births are phenylketonuric. These are the recessive homozygotes since the condition is produced by a single recessive gene. Following the procedures outlined in section 3.1 we can conclude that, if the population can be regarded as one in which mating is at random and if the frequency of the recessive homozygote is 0.0001, the frequency of the mutant allele, q, is $\sqrt{0.0001}$. Thus q=0.01, or 1%. It is therefore apparent that the mutant allele is far more likely to be present in the heterozygous condition than in the homozygous; in fact using the Hardy–Weinberg Law the expected proportion of heterozygotes would be 2pq, in this case 0.02. The recessive homozygote contains two copies of the mutant allele and the heterozygote only one so that distribution between heterozygotes and homozygotes occurs in the ratio 0.02/(2 × 0.0001)=100. That such an enormous proportion of the mutant alleles in the population occur in 'carrier' heterozygotes means that selection can eliminate very few in each generation, whether this selection be natural, because the severe mental retardation leads to institutionalization and very little reproduction, or artificial, because of some eugenic programme. It should be pointed out that such a eugenic programme does not exist in practice, although it has been suggested on a number of occasions that individuals affected by serious inherited diseases should be discouraged from reproducing.

Phenylketonurea will have natural selection acting against the allele responsible, and its existence in the population at a frequency of rather under 1% is presumably due to recurrent mutation. The most serious aspect of the disease is the mental retardation, but this can be prevented by careful dietary control if the condition is detected early enough. New born babies are now routinely screened by a test for the presence of large quantities of phenylpyruric acid in the urine and those diagnosed as positive are prescribed a diet which is very low in phenylalanine. Individuals on a low phenylalanine diet then develop quite normally except for a restriction on what they can eat. But this presumably removes the effect of natural selection against the allele and one side of the balance of forces which maintained an equilibrium with a frequency of about 1%. Since the other force, mutation, presumably remains active will not the frequency now steadily increase? The answers is that it will, but only very slowly for mutation rates are very low. If mutation is the sole remaining cause of change then the rate of increase of the allele will be proportional to the mutation rate. Thus a mutation rate of 10^{-6} would require a time scale of more than a million generations before it largely replaced the previous common allele, and even this would assume that mutation occurs in only one direction, and this is known not to be the case. Nevertheless the allele which has previously been rare will become much more common over a long time. Whether such a change represents a genetic deterioration in the human stock depends on an assessment of the personal and social 'costs' involved. In the case of phenylketonurea the individual loses some freedom of choice of food but the social cost is negligible. But for other diseases the treatment may be such that the individual's quality of life is considerably affected. It is now known that the disease is due to an inability to produce an enzyme called phenylalanine hydroxylase which catalyses the conversion of phenylalaline into another amino acid tyrosine. If this reaction cannot occur the concentration of phenylalanine builds up and is converted into phenylpyruric acid and other derivatives instead. It is probably these abnormal derivatives rather than phenylalanine itself which interfere with brain development. But if the treatment is the difference between life and death the individual will no doubt be prepared to put up with a great deal of inconvenience and hardship. As Maurice Chevalier once said about growing old: '. . . the only alternative is too awful to contemplate'.

Heritable diseases such as phenylketonurea where there is a defect in some aspect of the normal biochemical machinery are often called 'inborn errors of metabolism', and there is little doubt that an increase in the frequencies of the mutant alleles responsible for such changes represents an undesirable genetic change in human populations. There is one other category of genetic change which is usually included in the claim that the human genetic stock is deteriorating, and that is a supposed loss of resistance to infectious diseases following the use of techniques to reduce the incidence of the pathogen. That such changes do occur can be seen by considering sickle cell anaemia again. The frequency of the sickle cell gene is now much lower in populations of North American negroes. Even allowing for the probable effects of dilution of this frequency by intermarriage with North American whites, among whom the gene has a very low frequency, it remains clear that the gene has declined in frequency since these particular negro populations moved from West Africa where malaria was very common to North America where it has been rare. But in what sense is such a change undesirable? Malaria remains a very important disease on a world scale, but while there are serious problems about controlling the disease in those areas where it is endemic there does not seem to be a real risk that it will spread to other regions such as the U.S. And where the disease does not occur there is no doubt but that the population is better off without the sickle cell mutant for the homozygote condition is usually lethal.

We do not know how often a degree of resistance to, or tolerance of, an infectious disease or of some other environmental hazard has been achieved at the cost of carrying a concomitant disadvantage. But where it has, the manipulation of the environment to remove the particular environmental pressure will lead to an improvement of the genetic stock rather than a deterioration, for natural selection will gradually eliminate the associated disadvantage.

Thus improvements in public health measures, applications of medical science and other ameliorations of our environment may lead to genetic changes in human populations, but these will be rapid only if the modification has allowed natural selection to act against deleterious effects which had previously been balanced by some advantageous properties. If the alteration of the environment leads to the complete removal of selection on some variation the genetic changes will be very slow being of the order of the mutation rate. In this latter category will be some changes which will in some aspects be undesirable.

If, therefore, there is no great likelihood that any rapid deterioration of the genetic stock of man will occur, is there no cause for concern? Is eugenics irrelevant as well as possibly dangerous? We have seen that deleterious genes held in check by natural selection against recurrent mutation will be rare and that selection against them will, if they are recessive, be ineffective because they are rare. Thus any eugenic programme aimed at reducing the frequency of such genes would achieve nothing, especially if it were operated with the most humane motives that while sufferers from hereditary diseases should be treated to make their lives as normal as possible, they should be discouraged from breeding.

Are there then any characteristics which might be improved? It might be thought desirable, for example, to increase the average intelligence of a population or of the level of philanthropy. On this topic two comments should be made. Firstly it would be very difficult to reach agreement on the definition of the desired criteria; what kind of tests should be applied in order to measure intelligence or philanthropy? Secondly, the inheritance of such complex characters is itself very complex and only imperfectly understood even in the best studied example, which is performance in certain standard I.Q. tests. The main problem is that environmental variation has such an influence that it is extremely difficult to disentangle genetic and environmental effects on variation in I.Q. Thus the responses to selection are likely to be very slight. These arguments which suggest that any eugenic programme to improve a feature such as intelligence would be ineffective do not include the most compelling reason of all for not embarking on such a programme, that such a programme would be unacceptable to the population upon which it was imposed.

Does this lead to the conclusion that genetics has no contribution to make to human evolution? The answer is that it has if we turn our attention from gene frequencies and populations to a consideration of genotypes and individuals. The occurrence of a case of serious heritable affliction is a personal tragedy for the individual and for his parents. Where families at risk have been identified they can be offered advice about the likelihood of a child being born with a particular disease. This process of detection and the giving of information on which a couple can base a decision about having a child is called genetic counselling. All too often the detection of a family at risk depends upon the birth of a child with a serious disease, thus, in the case of a recessive character, identifying the couple as a pair of heterozygotes with the probability of producing a second such child being 25%. To push back the prediction to an earlier stage before any children are born requires either a great deal of data in the form of pedigree information for both partners or the existence of tests, probably biochemical, which can identify heterozygotes.

It should be stressed that the aim of genetic counselling is to offer information to couples who then make their own decisions; genetic counselling is not a part of any eugenic programme. Nevertheless some may see sinister implications in the storage of pedigree information in computer data banks, a development which would enormously facilitate the calculations of the risks involved for any particular couple before they have any children. Is it not possible that such information might be misused at some future time with a less democratic regime imposing pressure or coercion on certain couples not to marry or not to have children? It is important to keep this in proportion by reminding ourselves that the information to be recorded concerns characters which are serious heritable defects, such as muscular dystrophy, or spina bifida, and it would seem that the best guarantee that only such data are entered into the store is that it should be done by the medical profession rather than by a government agency. In this way genetic counselling will become a part of preventive medicine.

References

ALLISON, A. C. (1964). Polymorphism and natural selection in human populations. *Cold Spring Harb. Symp.quant. Biol.,* **21**, 137–149.

BABCOCK, E. B. (1947). *The Genus Crepis.* Univ. California Press, Berkeley and Los Angeles.

BAKER, H. G. (1964). *Plants and Civilisation.* Macmillan, London.

BOYD, M. F. (1949). *Malariology.* W. B. Saunders and Co., Philadelphia.

BRADSHAW, A. D., McNEILLY, T. S. and GREGORY, R. P. G. (1965). Industrialisation, evolution and development of heavy metal tolerances in plants. *Symp. Br. Ecol. Soc.,* **5**, 327–43.

CAVALLI–SFORZA, L. L. and BODMER, W. F. (1971). *The Genetics of Human Populations.* Freeman, San Francisco.

DARLINGTON, C. D. (1958). *Evolution of Genetic Systems.* Oliver and Boyd, Edinburgh.

DARWIN, C. (1899). *The Variation of Animals and Plants under Domestication.* John Murray, London.

DICKERSON, R. E. (1972). The structure and history of an ancient protein. *Scient. Am.,* **226**, (4) 58–72.

DOBZHANSKY, Th. (1970). *Genetics of the Evolutionary Process.* Columbia Univ. Press, New York.

DOBZHANSKY, T., AYALA, F. J., STEBBINS, G. L. and VALENTINE, J. W. (1977). *Evolution.* Freeman, San Francisco.

FORD, E. B. (1971). *Ecological Genetics.* Chapman and Hall, London.

GRANT, V. (1963). *The Origin of Adaptations.* Columbia Univ. Press, New York.

HALDANE, J. B. S. (1924). A mathematical theory of natural and artificial selection. *Proc. Camb. phil. Soc.,* **23**, 19–41.

KETTLEWELL, H. B. D. (1956). Further selection experiments on industrial melanism in the Lepidoptera. *Heredity,* **10**, 287–301.

KETTLEWELL, H. B. D. (1958). A survey of the frequencies of *Biston betularia* (L) (LEP) and its melanic forms in Great Britain. *Heredity,* **12**, 51–72.

KOEHN, R. K. (1969). Esterase heterogeneity: dynamics of a polymorphism. *Science,* **163**, 943–4.

KOEHN, R. K. and RASMUSSEN, D. I. (1967). Polymorphic and monomorphic serum esterase heterogeneity in Catostomid fish populations. *Biochem. Genet.,* **1**, 131–44.

KOESTLER, A. (1971). *The Case of the Midwife Toad.* Hutchinson, London.

LACK, D. (1948). Selection and family size in starlings. *Evolution,* **2**, 95–110.

MATHER, K. (1941). Variation and selection of polygenic characters. *J. Genet.,* **41**, 159–93.

MAYR, E. (1942). *Systematics and the Origin of Species.* Columbia Univ. Press, New York.

PATERNIANI, E. (1969). Selection for reproductive isolation between two populations of maize, *Zea mays. Evolution,* **23**, 534–47.

STRICKBERGER, M. W. (1976). *Genetics* (2nd ed.). Macmillan, New York and London.

RIGBY, P. W. J., BURLEIGH, B. D. and HARTLEY, B. J. (1974). Gene duplication in experimental enzyme evolution. *Nature,* **251**, 200–4.

WADDINGTON, C. H. (1953). Genetic assimilation of an acquired character. *Evolution,* **7**, 118–26.